Rheology
for Ceramists

Dennis R. Dinger

Rheology for Ceramists

Copyright © 2010
Dennis R Dinger
Second Edition

ISBN 978-0-557-81405-3

Published by: C B Dinger

Table of Contents

Preface

The main goal of this, the second volume in my series for ceramists, has been to produce a small reference book that explains the fundamental rheological topics that every ceramist should understand.

During my travels, I have seen far too many systems in plants that were well-designed for simple fluids, but that were being used to transport suspensions. In many ways, the viscous behaviors of suspensions are very similar to the viscous behaviors of simple fluids. But there are equally many ways in which the behaviors of suspensions differ markedly from the behaviors of simple fluids.

The most important of these differences is that suspensions contain particles. Of course, this is obvious! But suspended particles can settle; they can collide with one another; they can abrade the insides of pumps and pipes; they can pile up and cause problems; they can break and change size during flow; etc. None of these phenomena can possibly occur when the pipe train handles only simple fluids. It is absolutely necessary to consider all of these phenomena when the pipe train will handle suspensions.

Then, in addition to all of these suspension-specific considerations, ceramists must deal with the complex rheologies that are common in suspensions but which hardly ever occur in simple fluids.

Complex rheologies, which characterize suspensions, are treated as advanced subjects in most college curricula. Only in ceramics curricula are rheological properties normal topics for introductory processing courses.

One problem with the subject of rheology is that most textbooks are written for mathematical geniuses. Calculus and differential equations are subjects that most ceramists only take because they are required courses. This applies to me, too. So when I pick up a textbook to try to learn something about rheology, and I randomly open the book and am confronted with 42 differential equations on each page, I immediately close the book and return it to its shelf.

I know there is a need for a textbook that explains rheology in simple terms. I have tried to write that textbook.

As a youngster, I spent a lot of time with my Pennsylvania Dutch grandfather, and I learned to speak English with a very heavy accent. Since then, I have gotten rid of most of the heavy accent (my wife might disagree??), I have written a lot, and I have improved my English dramatically (I think so anyway). The last word anyone would ever use to describe my writing is *highbrow*. I don't use $10 words often in my writing because I don't know that many of them, and I certainly don't use them in my everyday speech.

In the preface to our ceramic processing textbook, Jim Funk said that we wrote it slowly because we know ceramic engineers read slowly. Well, I wrote this book slowly, too, and I gave it my best shot at making everything as clear as possible.

My goal for this book was to make it understandable to all ceramists: students, technicians, engineers, managers, artists, ... whomever. If I have successfully accomplished this, I will be very pleased. If not, I'm sure I'll hear about it.

I hope this book helps all who read it to understand some of the complexities of the rheology of suspensions. Hopefully, you will all also learn what I have: Suspension rheology is simply a fascinating subject.

Dennis R. Dinger
12 November 2002

Chapter One

Introduction

Rheology is the study of the viscous behaviors of fluids, suspensions, and forming bodies that occur over the full range of applied shear conditions. All phenomena pertaining to the deformation and flow of matter fall within this science called *rheology*.[1]

The complete range of shear conditions includes all possible shear rates from extremely low to extremely high values. An example of very mild, low shear conditions is the shear achieved when slowly stirring a glass of water with a spoon. An example at the high shear extreme is the intense shear applied to water as it passes through the nozzle of a garden hose. The shear in the nozzle is great enough to cause the bulk water in the hose to break down into a spray of tiny droplets as it exits the nozzle.

Viscosity

The *viscosity* of a fluid characterizes how easily it will flow when sheared. Water, gasoline, and paint thinner, for example, have low viscosities. When poured onto a sloped surface, each of these fluids will quickly flow down the surface. In contrast, molasses and cooking oils have higher viscosities. When poured onto the same sloped surface, they will also flow down the surface, but they will do so more slowly than the low viscosity fluids. We associate *low* viscosities with thin fluids that flow quickly, and we associate *high* viscosities with thicker fluids that flow more slowly.

The viscosity of water, which is a simple fluid, is constant. Even under the two very different shear conditions in the examples above (stirring and atomizing through a spray nozzle), water's viscosity is constant. It does not change as applied shear conditions vary. Constant viscosities are typical of simple fluids.

1

Some fluids, however, do not exhibit constant viscosity as shear rates vary. Unlike water, such fluids can have viscosities that vary widely as shear conditions change. When viscosities change as functions of shear conditions, the fluids are characterized by more complex rheologies and they are not simple fluids.

The field of *rheology* characterizes and quantifies the many possible viscous behaviors of the wide variety of existing fluids, suspensions, and forming bodies.

Viscosity vs Rheology

Rheology is a much broader subject than *viscosity*. *Rheology* is the parent subject which contains *viscosity*. In most common, simple fluids, of which water is the primary example, viscosity is constant at all shear rates. When a fluid is characterized by a single, constant viscosity over the whole range of shear rates, it is known as a *Newtonian* fluid. That is, it exhibits *Newtonian* rheology.

When dealing with simple Newtonian fluids, the magnitude of the imposed shear seldom enters into discussions of viscosity. Each Newtonian fluid is characterized by a constant viscosity; Newtonian viscosities do not vary as the magnitude of applied shear varies; so there's little need to be concerned with the magnitude of the applied shear. As a result, the rate of applied shear (i.e., the shear rate) is not prerequisite to defining the viscosity of Newtonian fluids.

Many people are not familiar with the terms *shear rate* and *rheology* for two reasons: (1) Most common fluids **are** Newtonian. (2) These two terms are not generally used in discussions of Newtonian fluids.

Many handbooks which contain viscosity tables list the viscosities of common fluids. If there are no accompanying explanations to describe the shear rate(s) at which the tabulated viscosities apply, the fluids described are Newtonian. If there are no labels on any of the fluids to indicate other rheologies, they are Newtonian. Tables like this are common in handbooks. They usually show fluid names, measurement temperatures, and viscosity values. All fluids in such tables are Newtonian, and the tabulated viscosities apply at all shear rates.

But many fluids exhibit viscosities that are NOT constant with shear rate. Such fluids are known as *non-Newtonian* fluids.

When fluids in a data table are non-Newtonian, the shear rate at which the measurement was made will accompany each non-Newtonian viscosity. The shear rate **must** be given or the data is meaningless. It would be useful (but it's not absolutely necessary) for such tables to also include the name of the type of non-Newtonian rheology exhibited.

Newtonian behavior is only one of the existing rheological categories of fluids. There are several different types of behaviors that fall into the broad category known as *Non-Newtonian* rheologies.

For example, fluids with viscosities that increase as shear rates increase exhibit *dilatant* rheology. Fluids which decrease in viscosity as shear rates increase exhibit *pseudoplastic* (a.k.a. *shear-thinning*) rheology. Suspensions which gel and exhibit yield stresses can be *yield-dilatant*, *Bingham*, or *yield-pseudoplastic* rheologies. Suspensions which decrease or increase in viscosity with time at constant shear rate can be *thixotropic* or *rheopectic* rheologies, respectively.

The *rheology* of a fluid defines how the viscosity behaves as a function of applied shear. The *viscosity* of a fluid defines how the shear stress varies as a function of applied shear. Each fluid exhibits a single, characteristic *time-independent* rheology. Non-Newtonian fluids can also simultaneously exhibit *time-dependent* rheologies. Depending on their characteristic rheological behavior, each fluid can exhibit a variety of measured viscosities as shear rates vary and/or as they are exposed to constant shear rates for periods of time. The field of rheology covers all of these interesting possibilities.

Rheology and Ceramists

Within most university curricula, *rheology* is an advanced topic studied by students only after they are well versed in the fundamentals of fluid mechanics and simple fluids. Most ceramists (engineers, artists, technicians, and managers), however, are introduced to the intricacies and complexities of rheology during their very first ceramics lab as a student, or during their very first day on the job in a ceramic production plant.

Typical ceramic bodies include both low solids content particle/fluid suspensions (known as *slips* and *slurries*) and high solids content particle/fluid plastic forming bodies. *Slurries* are suspensions of single ingredients. A ball clay slurry is a suspension that contains only ball clay. *Slips* are mixtures of ingredients. A body slip is a suspension that contains all body ingredients.

The fact that ceramic bodies are complex particulate suspensions (i.e., that they are multi-phase mixtures rather than simple fluids) brings all of the non-Newtonian rheologies into the realm of probability. That is, non-Newtonian rheologies are not just possible in ceramic suspensions, but probable.

This makes it necessary for ceramists to understand not only the subject of viscosity, but also the broader subject of rheology. In particular, ceramists need to understand the details of the cause-and-effect relationships that govern rheological and forming properties of production slips and bodies. They not only need to know **what** these behaviors are, but **why** they happen, and **how** to control them.

Rheologies

Simple Newtonian Fluids

Everyone is familiar with *simple* fluids because water is a simple fluid and everyone is familiar with water and its behavior. Simple fluids each consist primarily of large volumes of a single molecular type, although simple fluids can include mixtures of molecular types. Water, most low viscosity cooking and fuel oils, and many solvents are examples of simple fluids. In most cases, simple fluids are *Newtonian*, because they each exhibit a characteristic, constant viscosity, regardless of shear conditions.

When someone asks, "What is *the* viscosity of water?" or "What is *the* viscosity of vegetable oil?", they are announcing (whether they know it or not) that the fluid viscosity in each case will be a single, constant value and that the fluids in question are Newtonian fluids.

Most people are familiar with the concept of *viscosity* because they have had first hand, practical experience with simple fluids. They can recognize and distinguish between high and low viscosity fluids. But they do not necessarily know that simple fluids are *Newtonian* fluids, nor that other more complex fluids within their experience are known as *non-Newtonian* fluids.

Non-Newtonian Fluids

Fluids that do not exhibit constant viscosity as shear rates vary are called *non-Newtonian* fluids. As a rule, non-Newtonian behaviors occur in multi-phase fluid mixtures and in mixtures of immiscible liquids. Liquids mixed with fine particles (particle/fluid suspensions), foams (liquids containing large volumes of gas bubbles), and three-phase mixtures of solid particles, liquids, and gas bubbles are examples of multi-phase mixtures.

Any time one works with complex, multi-phase mixtures, the assumption should be made that the mixtures will exhibit non-Newtonian rheological properties.

Shear-Thinning (Pseudoplastic) Behavior

Most people **are** familiar with non-Newtonian fluids, even though they may not be familiar with the label *non-Newtonian*. At one time or another, most people have successfully (or unsuccessfully) tried to pour ketchup from a bottle. Most ketchups exhibit the non-Newtonian rheology known as *shear-thinning* rheology. A full bottle of ketchup flows poorly (or not at all) until the ketchup is subjected to sufficient shear to break up its gel structure, reduce its viscosity, and allow it to flow.

How many of us have splattered ketchup all over the dining room table in an attempt to put a little bit onto a hamburger? Shaking the ketchup bottle and hammering on its bottom with one's fist usually increases in intensity until the ketchup begins to flow. By the time that happens, however, many of us are thoroughly frustrated and are beating and shaking the bottle violently. Once the yield stress has been exceeded, flow begins, viscosity drops quickly, and the ketchup flows quite well. Under such violent conditions, ketchup not only flows well but it will usually splatter all over everything within about a 3' radius of the target. Ketchup is a *shear-thinning* non-Newtonian fluid and these are typical shear-thinning properties.

Have you ever wondered how viscous a fluid must be to support a serving spoon that has been placed into it? Whipped cream can do this. Mayonnaise can do this as well. Whipped cream and mayonnaise can each support a spoon in whatever position it is placed, even if the spoon doesn't touch the container. How viscous are whipped cream or mayonnaise?

Water can't do this. Neither can coffee. A spoon placed into a cup of coffee or water will fall to the bottom of the cup, and then the spoon's handle will drop against the lip of the cup.

A log can support an axe or a hunting knife that has been plunged into it. But a log is a solid, not a fluid, nor a suspension.

If whipped cream and mayonnaise were Newtonian fluids, the following questions show the dilemma: If these fluids are viscous enough to support a spoon as just described, how much force must be applied to the spoon to insert it into the whipped cream or mayonnaise? ... or to move it? ... or to serve the whipped cream onto dessert? ... or to serve the mayonnaise into a salad?

If whipped cream and mayonnaise were Newtonian fluids, their viscosities would necessarily be extremely high to support the spoon as described. In fact, their viscosities would be so high that they would both be considered solid materials. To insert the spoon into a solid, or to stir, scoop, or serve such a material with the spoon would be impossible.

Fortunately, whipped cream and mayonnaise are not simple Newtonian fluids, nor are they solids. They are both non-Newtonian fluids. The force that has to be applied to the spoon to stir either fluid is not proportional to the force that supports the stationary spoon.

Another test to perform is this: Remove a container of ketchup, whipped cream, or mayonnaise from the refrigerator and carefully lay it on its side on the kitchen counter. If this is done with sufficient care, the surface of the ketchup, whipped cream, or mayonnaise within the container will not move — even when the container is laying on its side and the surface is vertical. This behavior would not happen with water or other simple fluids.

Ketchup, whipped cream, and mayonnaise are examples of shear-thinning materials which exhibit yield stresses. One must apply enough stress to each to exceed its yield stress before flow can occur. The yield stress prevents a fresh bottle of ketchup from flowing easily. The yield stress is strong enough to support the spoon in a bowl of whipped cream or in a jar of mayonnaise.

After the yield stress has been exceeded and flow has begun, each of these materials can be stirred, whipped, and served with ease because they each have relatively low viscosities after their yield stresses are exceeded. Under shear, the structures that produce the yield stress are broken down, thereby allowing flow. When flow stops, the structures rebuild, the yield stresses reappear, and they can each again support a spoon.

Ketchup, mayonnaise, and whipped cream are all examples of shear-thinning fluids stored in most home refrigerators. Everyone who has ever eaten any of these has had practical experience dealing with non-Newtonian, shear-thinning fluids.

Dilatant Behavior

The most common example at the opposite rheological extreme from shear-thinning behavior is a mixture of corn starch and water which many chefs use to thicken gravy. Corn starch in water provides an excellent example of *dilatant* rheology, in which viscosities increase as shear increases. Dilatant fluids must be stirred and sheared slowly because the faster the stirring action (and the higher the level of imposed shear), the greater their effective viscosities, and the more difficult they are to mix.

Dilatancy in a production process can cause many severe problems. The solution to minimizing dilatant effects in a suspension known to be dilatant is to lower the rate of applied shear. The direction of this change is contrary to the way most of us think. To successfully mix a dilatant suspension, the mixing action must be gentle, low shear. It must not be increased to more intense, higher shear rates. Increasing the shear intensity to better mix a dilatant suspension simply will not work.

A ceramist who wants to work successfully with dilatant fluids faces a difficult, counter-intuitive problem. Everyone 'knows' when a suspension is thick and pasty, that one must put more muscle into the mixing operation for it to be successful. If such a suspension is to be mixed in a mixer, everyone 'knows' that you simply increase the speed of the mixer to increase the intensity of mixing. If the mixture is truly dilatant, however, the **only** solution that will work is to **reduce** the intensity of mixing – that is, to reduce the speed of the mixer. Increasing the applied stresses will work with high viscosity Newtonian fluids, and with shear-thinning fluids, but not with dilatant fluids.

When mixing dilatant suspensions, high shear, high intensity conditions will cause problems. Increasing the mixing intensity in a dilatant suspension improves the likelihood that the mixer motor will burn up, without providing any improvement to the homogeneity of the batch. In such cases, high shear conditions will produce less mixing action than low shear conditions (and maybe no mixing at all).

The main goal since the beginning of the industrial revolution has been to increase process speeds and reduce mixing times. The solution to successfully working with dilatant suspensions, however, is to reduce process speeds and to increase mixing durations. Management may not like this solution, but it is the correct and necessary solution to the dilatancy problem.

When working with dilatant systems, reducing shear intensities during mixing and during other process steps will reduce the deleterious effects of dilatancy. With reduced shear intensities and increased mixing times, such powders and fluids can successfully be well mixed. But good luck! Successfully dealing with dilatant fluids is much easier said than done.

The long range solution to a dilatancy problem is to change (eliminate) the conditions that caused the dilatancy in the first place. When the current production batch is dilatant (which may represent quite a few tons of body) and it must be used, process shear intensities must be reduced.

Multi-Phase Systems

Generally speaking, one should assume (until proven otherwise) that *any* **multi-phase** mixture **will** exhibit **non**-Newtonian properties. Non-Newtonian fluids are the norm in ceramic process systems. Non-Newtonian fluids are prevalent in many other industrial process systems as well.

Since almost all ceramic production bodies are multi-phase, complex suspensions which exhibit non-Newtonian rheologies, ceramists must be familiar with the subject of rheology.

Summary

In this chapter, we called upon the practical experience that everyone has had with common fluids to introduce the concepts of viscosity, rheology, Newtonian fluids, and non-Newtonian fluids.

Most of us, especially those who are practicing ceramists, deal with non-Newtonian fluids on a daily basis. All who work with such fluids and suspensions, and especially those who are responsible for production and control of non-Newtonian slips and forming bodies, need to be familiar with the intricacies of non-Newtonian phenomena, as well as the cause-and-effect relationships that allow the modification and control of those phenomena. Discussions of all such topics will follow in later chapters of this textbook.

Chapter Two

Fundamentals

In this chapter, the two rheological terms *shear rate* and *shear stress* will be defined and discussed. Then, they will be used to define the concept of *viscosity*.

Shear Stress

A stress is a force applied over an area. *Shear stresses* are stresses that result from the application of opposing forces that are parallel to one another, but offset so they are not in line with one another. The resulting stress shears the object. It doesn't put the object under tension or compression, but under shear.

Figure 2.1 shows tensile, compressive, and shear forces acting on an object. The solid lines show the objects before the application of the forces. The dashed lines show the types of deformation caused by the applied forces.

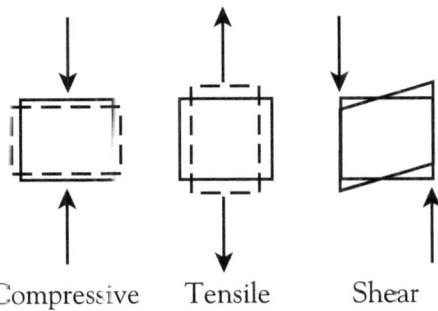

Compressive   Tensile   Shear

Figure 2.1 Compressive, Tensile, and Shear Forces

Note that the compressive and tensile *forces*, when applied uniformly to the top and bottom surfaces of cubes, become compressive and tensile *stresses*, respectively. Compressive stresses squeeze a cube. Tensile stresses stretch a cube.

The shear forces, however, don't act on the upper and lower faces of the cube. They act on opposing side faces of a cube. When applied uniformly across the left and right faces of the cube, they create shear stresses. As the stresses shear the cube, they deform it as shown.

When stresses are applied to elastic solids, the solids are compressed, elongated, or sheared. When the stresses are removed from elastic solids, the elastic deformations are relieved, and the objects revert back to their original shapes.

When shear stresses are applied to fluids, however, individual layers of fluid molecules move relative to one another. When shear stresses are removed, the fluid molecules stop in their new positions, and the fluid holds its new molecular arrangement. Flow occurs within simple fluids as soon as shear stresses are applied.

Simple fluids, which do not exhibit yield stresses, will flow under any applied shear stress. The rate of shear achieved will be a function of the viscosity of the fluid. Even the smallest shear stress will cause a simple fluid to flow, and its molecules to rearrange.

Those fluids which exhibit yield stresses will only begin to flow after the applied shear stress exceeds the yield stress. Up to that point, the fluid structures may undergo elastic deformation, but after the yield stresses are exceeded, molecules will rearrange, and flow will occur.

Ceramic process suspensions are known as visco-elastic materials. This refers to the fact that they exhibit both elastic and viscous properties.

Practical Application of Shear Stresses

Figure 2.2 shows the fundamental diagram used to define and explain the terms *shear stress*, *shear rate*, and *viscosity*. It depicts two parallel plates with a fluid contained between them.

When a force to the right is applied to the upper plate, and an equal and opposite force to the left is applied to the lower plate, the fluid between the two plates is subjected to shear. The forces applied uniformly across the areas

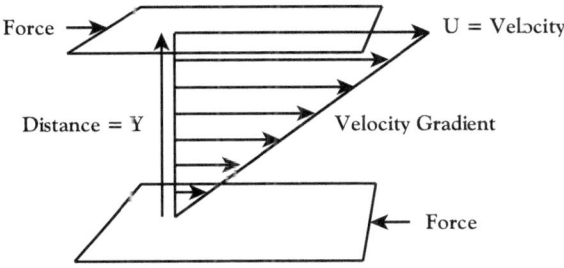

Figure 2.2 The Fundamental Definition of Viscosity

of the plates produce the shear stress and the plates in contact with the fluid transmit the shear stresses to the fluid.

The shear stress in this diagram is:

$$\text{Shear Stress} = \frac{\text{Force}}{\text{Area}} = \tau_s = \frac{F}{A} \ [=] \ \frac{N}{m^2} = \text{Pa} \qquad (2\text{-}1)$$

Note that the dimensions of shear stress are force/area, which are commonly associated with pressure units. In engineering units, the *psi* is the common unit for shear stress. In SI units, the *Pascal* is the common unit.

In concentric-cylinder viscometers, and in some process devices, shear stresses are applied to fluids between one moving and one fixed surface. In some concentric-cylinder (cup-and-bob) viscometers, the motor causes the bob to rotate and to shear the fluid against a stationary cup. This is the case in Figure 2.2 when the upper plate moves to the right and the lower plate remains stationary.

In some viscometers, the motor drives the bob at a particular rpm, and it is coupled to the bob through a spring sensor which can measure the instantaneous force on the spring. Since the surface areas of the outside surface of the bob and the inside surface of the cup are known, and the applied force can be measured, the shear stresses applied to the cup and bob, which are transmitted across the fluid, can be calculated.

In some processes, such as extrusion, the outer (circumferential) surface of the auger moves relative to the fixed barrel of the extruder.

Admittedly, this is not the major force applied by the auger to the extrusion body, but it is a shear force that shears that part of the body that happens to lie between the outer edge of the auger and the barrel.

Distances between the auger and the barrel are usually small, so only small volumes of body are affected. But the auger/barrel gap in an extruder, or the corresponding impeller/housing gap in a pump, are regions within these devices that impart relatively high shear to fluids, suspensions, and bodies. The bulk of the fluids, suspensions, or bodies within such devices is exposed to low shear and to compression, but the magnitudes of those stresses are much more difficult to calculate or estimate.

Pumps push fluids and suspensions through piping systems, so the shear stresses in pipes show themselves as drag between the flowing fluids and the stationary pipe walls. Flowing fluids and suspensions in pipes have velocity profiles from wall to wall within the pipe. The fastest flow rates are at the center of the flow channel and the slowest flow rates are near the walls.

Many teach that the layer of fluid in contact with the pipe wall remains stationary relative to the wall. This appears to be true for the layer of molecules of simple fluid in contact with a pipe wall. It may even be true for the layer of molecules of carrier fluid in a suspension in contact with a pipe wall. But it does not appear to be true for suspended particles in the vicinity of the pipe wall.

During pipe flow of a simple fluid, flow rates will vary from the maximum at the center of the pipe to zero at the walls. During pipe flow of a particle/fluid suspension, flow rates will vary from the maximum at the center of the pipe to low, but possibly non-zero, values at the walls.

When suspended particles stop and remain stationary at the wall, blockages can form and process problems can occur. This is a more complex issue that will be dealt with later.

Shear Rate

When a shear stress is applied to a fluid, the fluid will deform at a steady rate. This rate of deformation is known as the *shear rate*. The shear rate is equal to the value of the velocity gradient of the molecules and particles at the point of interest within the fluid. The diagram in Figure 2.2 depicts the shearing of a simple fluid between two parallel plates. The shear rate is equal to the velocity gradient shown in that diagram.

For the geometry shown in Figure 2.2, the velocity gradient in a simple fluid is linear. That is, the shear rate of the simple fluid in Figure 2.2 is constant at all points between the two plates. The velocity gradient shown in Figure 2.2 does not apply to non-Newtonian fluids, however, because the velocity gradient between two plates is non-linear for non-Newtonian fluids. A linear velocity gradient applies only to simple fluids.

To calculate shear rate, one only needs to calculate the value of the velocity gradient at the point of interest. Since the velocity gradient in this figure is linear, the calculated shear rate applies at any point between the two plates.

$$\text{Shear Rate} = \frac{\text{Change in Velocity}}{\text{Distance}} = \frac{\Delta U}{\Delta Y} = \dot{\gamma}$$

$$\dot{\gamma} \; [=] \; \frac{\text{cm/sec}}{\text{cm}} = \frac{1}{\text{sec}} = \frac{1}{s} = s^{-1}$$

(2-2)

Note that the shear rate, $\dot{\gamma}$ [gamma dot], is the velocity gradient with units for velocity/distance, or (cm/sec)/cm, as shown. Shear rates are velocity gradients. They should be remembered as such. Units for velocity gradients are velocity/distance: (cm/s)/cm, (m/s)/m, or (ft/s)/ft, etc.

We tend to simplify everything to its extreme, so shear rate usually carries the units s^{-1}, that is, reciprocal seconds (1/s). What is the practical meaning of a reciprocal second? What is the meaning of a velocity gradient? A velocity gradient has a practical, understandable meaning. A reciprocal second does not. Throughout this book, when the term *shear rate* is used, even if its units are given as s^{-1}, the words that should come to mind are *velocity gradient* and the units that should come to mind are *(cm/s)/cm* or another appropriate complete set of units for velocity gradient

Calculating or Estimating Shear Rates

How does one calculate or estimate shear rates? When there are two moving surfaces, such as there are in Figure 2.2, it's relatively easy to perform the calculation. When the fluids are non-Newtonian, the calculation should initially be performed as if the fluids were Newtonian. Then, the calculated shear rate value can be increased or decreased as necessary to accommodate the non-Newtonian fluid.

When dealing with pipe flow, where suspensions move through pipes, the radius of the pipe is the distance over which the estimate should be made. The fastest flow velocity is usually at the center of the pipe and the slowest at the wall. The radius defines this distance.

When suspensions flow through channels with rectangular cross-sections, which are somewhat similar to Figure 2.2 but with the suspension flowing and both plates stationary, again the shear rate should be calculated from the center of the flow channel to the nearest wall.

For fluid flow within a fixed channel or pipe, one also needs to know the maximum flow velocity of the fluid. The average flow rate is fairly easy to measure using a volumetric flow meter. Many different flow meters can be used, but a volumetric flow meter can be as simple as a bucket and a stopwatch. For atomizers, the total flow rate through the atomizer divided by the number of individual channels supplying the flow, will produce the average flow rate per channel.

The average flow rate (volume/time) divided by the cross-sectional area of the channel (area) produces the average flow velocity:

$$\frac{\text{volume/time}}{\text{area}} = \frac{\text{length}^3/\text{time}}{\text{length}^2} = \frac{\text{length}}{\text{time}} = \text{velocity} \qquad (2\text{-}3)$$

The maximum flow velocity is not easy to measure. As a starting point, the maximum flow velocity can be assumed to be twice the average value. It may not be exactly correct, but it will be close. When calculating shear rates, the exact value of the number is not as important as the order of magnitude of the number. Whether the calculated shear rate is $1000s^{-1}$ or $2000s^{-1}$ or $1345.692s^{-1}$ is less important than knowing that the shear rate is in the 1000s, rather than in the 10s, 100s, or 10,000s.

Sample Shear Rate Calculation

If there are 50 channels 1/16" in diameter in an atomizer feeding 500 gph of fuel oil into a burner, what is the estimated shear rate imposed on the oil as it passes through each orifice?

$$\frac{500 \text{ gal}}{\text{hr}} \text{ X } \frac{3785\text{cm}^3}{1 \text{ gal}} \text{ X } \frac{1 \text{ hr}}{3600 \text{ s}} \text{ X } \frac{1}{50 \text{ channels}} = \frac{10.51 \text{ cm}^3/\text{s}}{\text{channel}} \qquad (2\text{-}4)$$

$$1/16" \text{ diameter} = 0.15875 \text{ cm diameter} \qquad (2\text{-}5)$$

The cross-sectional area of each 1/16" channel is:

$$\pi r^2 = 3.1416 \ (0.07938 \text{ cm})^2 = 0.01980 \text{ cm}^2 \qquad (2\text{-}6)$$

The average flow velocity in each channel is:

$$(10.51 \text{ cm}^3/\text{s}) \ / \ (0.01980 \text{ cm}^2 \ ) = 530.8 \text{ cm}/\text{s} \qquad (2\text{-}7)$$

A first estimate of the shear rate in each channel is equal to the velocity gradient between the flow velocity at the center of each channel and the wall (which is assumed to be zero). A good first estimate of the flow velocity at the center of the channel is twice the average flow velocity. The radius of each channel is the distance over which the velocity gradient is to be calculated. So the estimate is:

2 X (average flow velocity)/radius =

$$2 \text{ X } \ (530.8 \text{ cm/s}) \ / \ 0.07938 \text{ cm } = 13{,}370 \ (\text{cm/s})/\text{cm} \qquad (2\text{-}8)$$

$$= 13{,}370 \text{ s}^{-1}$$

Even if this estimate is off by a factor of 2 or 3, it nevertheless shows the order of magnitude of the shear rate. In particular, it emphasizes the point that shear rates in atomizers can be extremely high. One might also wonder whether the 1/16" channels are the smallest orifices in the atomizer? It is possible to learn, after disassembling the atomizer, that the fluid must pass through even smaller channels internally, or that it must pass through slightly larger orifices that are fewer in number as it flows through the atomizer. The most severe shear rate in an atomizer, however, should be at (or very near) the tip where the spray begins, so the values calculated in this example should be the highest shear rates imposed upon the fluid.

If the fuel oil is non-Newtonian, the highest imposed shear rate might be slightly higher than the value calculated, but the order of magnitude of the calculated value should be correct. The fuel oil passing through this atomizer is being subjected to extreme shear conditions as it exits the atomizer. High shear conditions are characteristic of atomizers.

Shear rates in holding tanks and other relatively quiescent systems are relatively low, on the order of 1 to $10s^{-1}$. Shear rates in piping systems are on the order of $10s^{-1}$ to $100s^{-1}$. Shear rates in pumps can be more intense ($100s^{-1}$ to $1,000s^{-1}$), and shear rates in atomizers will be even higher ($1,000s^{-1}$ to $10,000s^{-1}$).

Shear rates during some apparently slow operations such as brushing (painting) can be quite high due to the small gap between the brush and the surface.

Note that high shear rates are large velocity gradients. Velocity gradients can be large because the fluid flow velocities in a normally sized channel are high. When fluid flow velocities appear to be normal (or even low), velocity gradients can still be large because the size of the gap within which the shear takes place is very small.

When in doubt about a particular process, do a calculation to estimate the shear rate.

The Definition of Viscosity

Consider again Figure 2.2 which shows the relationship between shear stress and shear rate. Shear stresses are applied by the parallel plates to the fluid between the plates. At each applied shear stress, the system will come to equilibrium at a particular shear rate. The relationship between the applied shear stress, and the shear rate achieved, is defined by the viscosity of the fluid. Its defining equation is:

$$\text{Dynamic Viscosity} = \mu = \frac{\text{Shear Stress}}{\text{Shear Rate}} = \frac{\tau_s}{\dot{\gamma}}$$

$$\mu \ [=] \ \frac{\text{Pa}}{s^{-1}} = \text{Pa} \cdot s$$

(2-9)

The *dynamic viscosity*, μ, is the ratio between the shear stress and the shear rate. A high viscosity fluid will take considerably more applied stress to achieve a particular shear rate than will a low viscosity fluid.

Note that the *centiPoise* (cP), a common unit for viscosity, is equal to a *milliPascal second* (mPa · s):

$$1 \text{ cP} = 1 \text{ mPa} \cdot \text{s} \qquad (2\text{-}10)$$

The traditional way to measure dynamic viscosity, m, is to measure the shear stresses required to shear fluids at a variety of shear rates. Figure 2.3 shows a plot of measured shear stress versus measured shear rate. Each viscosity curve in this type of figure is

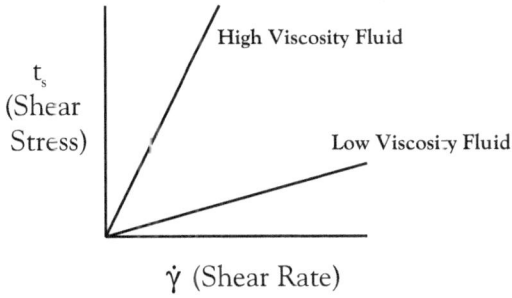

Figure 2.3 Rheogram of Two Simple Fluids

known as a *rheogram*. Figure 2.3 shows examples of both low and high viscosity fluids. These are Newtonian fluids because they exhibit linear behavior that begins at the origin (at zero shear stress and zero shear rate). The high viscosity fluid's rheogram has a steep slope. The low viscosity fluid's rheogram has a gentle slope.

The equations for these two fluids are the equations for straight lines with the Y-intercept equal to zero (B = 0). The equation for a straight line is:

$$Y = mX + B \qquad (2\text{-}11)$$

Substituting the shear stress, viscosity, and shear rate into their appropriate positions produces:

$$\tau_s = \mu \, \dot{\gamma} \qquad\qquad (2\text{-}12)$$

where τ_s = shear stress,
        $\dot{\gamma}$ = shear rate, and
        μ = dynamic viscosity.

Equation (2-12) is a rearrangement of Equation (2-9).

The viscosity of the high viscosity fluid in Figure 2.3, μ_H, which is the ratio of shear stresses to shear rates in the high viscosity rheogram, is equal to the slope of the rheogram as shown by Equations (2-11) and (2-12). Similarly, the viscosity of the low viscosity fluid, μ_L, which is the ratio of shear stresses to shear rates in the low viscosity rheogram, is equal to the slope of its rheogram.

Viscosities of both the high and low viscosity fluids in Figure 2.3 are constant values that do not change as shear rates change. Because their viscosities are constant, these two fluids are known as Newtonian fluids. It is important to note that the viscosities of Newtonian fluids are single, constant values that apply over the whole range of shear rates.

Calculating the ratios of shear stress to shear rate at each shear rate and plotting the values versus shear rate produces Figure 2.4. This type of plot clearly shows that Newtonian fluids have constant viscosities over the full

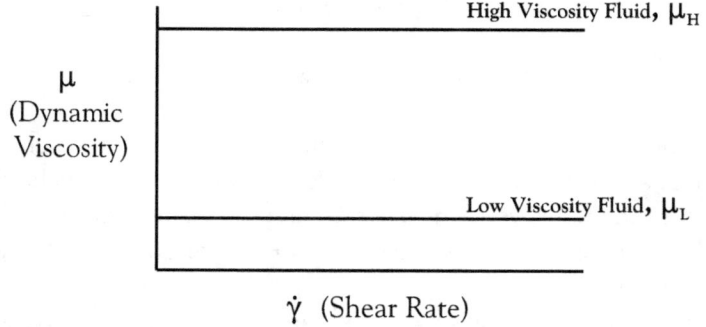

Figure 2.4 Dynamic Viscosity versus Shear Rate

range of shear rates. For this reason, when discussing Newtonian fluids, *'the'* viscosity of each fluid will frequently be mentioned. Newtonian fluids only have one viscosity, which is equal to the slope of the rheogram as shown in Equation (2-12).

The proper definition of *viscosity* for Newtonian fluids, and of *apparent viscosity* for non-Newtonian fluids, is: *viscosity is the ratio of shear stress to shear rate at any applied shear rate.* The instantaneous slope of non-Newtonian fluids at any shear rate is not equivalent to the viscosity of the fluid. The ratio of shear stress to shear rate at any imposed shear rate happens to be equal to the mathematical slope of the rheogram equation only for Newtonian fluids.

Viscometers that measure apparent viscosities typically measure the shear stress at each shear rate. Rheograms from such viscometers are usually plotted on shear stress versus shear rate axes, as in Figure 2.3. Some viscometers, which can calculate the apparent viscosity (the ratio of shear stress to shear rate) at each shear rate, can also plot rheograms similar to the one in Figure 2.4.

Summary

The discussion in this chapter covered shear stresses, shear rates, and viscosities. The simple relationship between the three is that the dynamic viscosity of any fluid under any set of conditions is the ratio of the measured shear stress to the measured shear rate.

Newtonian fluids are characterized by a single dynamic viscosity because the shear stresses required to achieve any particular shear rate are always proportional.

Non-Newtonian fluids, which will be discussed in more detail throughout this book, are characterized by apparent viscosities that differ at each shear rate.

Chapter Three

Simple Fluids versus Suspensions

The subject of fluid mechanics covers simple fluids under static and dynamic conditions. This text is not meant to replace any of the good fluid mechanics textbooks that are available today. In fact, ceramists would be wise to become familiar with the fundamentals and calculations that can be performed using fluid mechanics.

Most introductory fluid mechanics textbooks and courses, however, are restricted to simple, Newtonian fluids. Rheology and non-Newtonian behaviors are only covered in advanced courses.

Many engineers who are familiar with simple fluids do not appear to understand the differences between simple fluids and suspensions. This fact usually shows itself in the designs of suspension mixing, piping, and processing systems in plants.

The purpose of this chapter is to point out and discuss some of the fundamental differences between simple fluids and suspensions.

Suspensions

Although ceramists frequently use simple fluids, such as water, most ceramists must mix, modify, and control particle/fluid suspensions. Multiphase suspensions are not simple fluids.

A typical ceramic suspension is a mixture of fine powders with fluids (which could be water or non-aqueous fluids) in the presence of minor additions of a variety of inorganic and organic additives. Depending on the mixing equipment and procedures used, many ceramic suspensions also contain air bubbles.

The major contributors to the rheological behavior of suspensions are the properties and characteristics of the powders and of the carrier fluids. The additives, which usually comprise a fraction of a percent (by mass) of the powder in a suspension, modify and control the interfacial properties and interactions between the powders and the fluid, and thereby also modify and control the rheological properties of the suspensions.

Simple fluids do not have this wide range of variables to adjust and control to achieve process viscosities. Of course additives can be mixed with simple fluids to alter their behaviors, but the extent of changes possible to the viscous behaviors of simple fluids is minor by comparison to the extent of the changes possible in suspensions.

Two major phenomena that are not characteristic of simple fluids contribute to the behavior of suspensions: gelation and particle/particle interactions. Each of these will be discussed briefly in the following sections, and again in later chapters.

Powder Effects on Rheology

There are several categories of effects in which the powder fractions of suspensions play major roles in the control of rheological properties. The solids contents of suspensions, the particle size distributions of the powders, the properties of the powder constituents, the flocculation states of suspensions, and the nature and intensity of particle/particle collisions (which are closely associated with the imposed process shear rates) must each be considered. Then, because suspensions contain solid particles, unwanted settling, packing of particles, blockages, and abrasion must also be considered.

Because simple fluids contain no particulate solids, none of these considerations are applicable to simple fluids.

Solids Content

In suspensions, the percentage of solids has a major effect on rheological properties. When solids contents are low, fluid viscosities increase slowly as solids are added. Einstein's equation,[2] which applies to low solids content suspensions, describes this:

$$\mu_r = 1 + 2.5C \qquad (3\text{-}1)$$

where μ_r = relative viscosity, i.e., the suspension viscosity
relative to the viscosity of the carrier fluid, and
 C = the volume concentration (fraction) of particles.

At low solids contents when the volume concentration of particles, C, is small, suspension viscosity closely approximates the carrier fluid viscosity. In such cases, the fluid viscosity dominates to control suspension viscosity. As solids contents rise, suspension properties quickly deviate from their similarities to the properties of the carrier fluid.

At high solids contents, the particulate fraction has the major controlling effect on suspension viscosities. In high solids content suspensions, Einstein's equation does not apply. Each suspension will have an upper solids content limit, beyond which flow ceases. This occurs when a continuous network of touching particles spans the bulk volume of the suspension and particles are no longer free to move within the suspension. In each particular suspension, the limiting solids content depends upon the particle size distribution of the powder and the capability of those particles to pack. Some suspensions will remain fluid at higher solids contents than others due to particle size distribution and particle packing differences.

Figure 3.1 shows relative viscosities over the range of solids contents, taken from experimental and computer modelled data in the literature[3-6]. Note that Einstein's linear relationship, Equation (3-1), only applies at low solids contents with solids fractions $< \sim 0.3$-0.4. Particle/particle interactions quickly dominate viscous behavior as solids contents rise beyond that range. Figure 3.1 shows all viscosities rising exponentially as solids contents increase above $\sim 40\%$[3-6].

At high solids contents, too many particles and too little fluid are present to produce flow. Under high solids conditions when flow and deformation no longer occur, the term *suspension* no longer applies. Such systems behave like wet compacts of solid particles.

In the 1980s the author participated in a project where the goal was to make high solids, low viscosity suspensions. Some of the best, highest solids content suspensions produced during that project had solids contents greater than 70 vol% with viscosities less than ~ 400 cP.

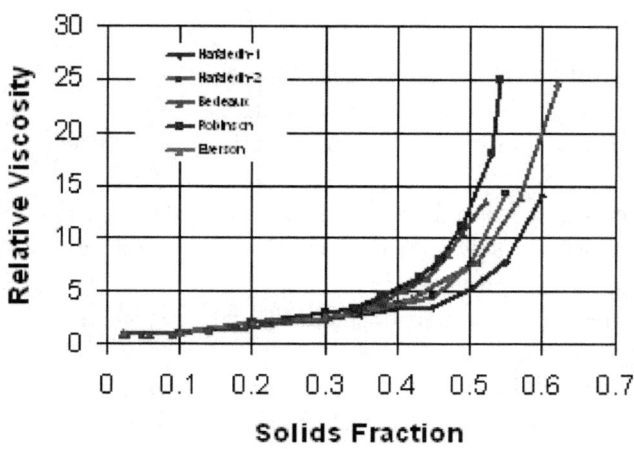

Figure 3.1 Relative Viscosity versus Solids Fraction[3-6]

At that same time, many coal storage piles at local power stations (huge piles of coarse chunks of coal) contained greater percentages of water than was present in the very fluid suspensions produced in our research project. To compact the coal storage piles at power stations, they routinely drive huge bulldozers across the tops of the piles. The particle size distributions in coal storage piles are very different from those that were in our suspensions. The particles in the our high solids content suspensions packed exceptionally well. The chunks of coal in storage piles don't pack well at all. Our suspensions were fluid (very fluid) at solids contents at which the coal piles exhibited no fluid properties at all.

Particle Settling

The sizes of particles in suspension affect the stability of suspensions. Colloidal particles will remain suspended within fluid systems even after long settling times. Large particles will settle quickly when suspensions are at rest and flow velocities are low.

The severity of settling depends primarily on the sizes of the suspended particles, but secondarily on the flow velocity of the suspension. If several bowling balls thrown into a swimming pool constitute a suspension, obviously the bowling balls will all settle quickly and rest on the bottom of the pool. Many fine powders in the same pool, however, will remain suspended and continue to circulate with the water until they come in contact with, and are removed by, the filter.

Using Stokes' law, the settling rates of particles in fluids can be calculated:

$$V = \frac{h}{t} = D^2 \frac{(\rho_P^2 - \rho_L^2)g}{18} \mu \qquad (3\text{-}2)$$

where V = velocity of a settling particle,
h = distance the particle settles in time t,
D = particle diameter,
ρ_P = particle density,
ρ_L = liquid medium density,
t = settling time,
g = gravitational constant, and
μ = liquid medium viscosity.

Stoke's law calculates 'unhindered' settling rates. Stoke's law does not apply when circumstances cause settling to be 'hindered.' One example of this is when too many particles are in suspension, and particles crowd one another. Another example is when huge particles settle and the turbulence in their wakes drags along other particles. Stoke's law doesn't apply in either of these cases.

Stoke's law also cannot be used to calculate the settling of colloidal particles. Colloids are affected by Brownian motion. They experience gravitational settling forces as do all particles, but Brownian motion can cause colloids to move randomly in any direction — up, down, or sideways. So the positions of colloidal particles are not predictable using Stoke's law.

Some mineral transportation companies pump aqueous suspensions containing relatively coarse particles from the mine sites through pipelines to their points of use. It is well known that large particles will settle quickly in these pipelines when pumping stops, such as during periods of maintenance or during unexpected breakdowns.

Each such pipeline, therefore, has a minimum acceptable flow velocity. Turbulence within the pipeline, at all velocities above the minimum, prevents particles from settling and keeps them entrained in the flowing stream. During maintenance periods, these pipelines must be emptied of their suspensions and filled with water. This requires that large holding tanks for both suspensions and water be placed at intervals along the length of the pipeline. Prior to maintenance periods, suspensions are emptied into the tanks, and water is pumped in to fill the pipeline.

These are extraordinary, but necessary, measures to take to use such pipelines. If those suspensions were ever allowed to settle in the pipelines, it would be difficult (and probably impossible) to ever get them flowing again and keep the pipelines free of obstructions.

Particle Size Distribution

The nature of the Particle Size Distribution (PSD), which refers to the whole distribution, not just the median particle size of the distribution, has a major effect on the rheological properties of suspensions. A PSD that packs well, and defines a small internal porosity when densely packed, can remain fluid to very high solids contents. A PSD that does not pack well will lose its fluid properties relatively quickly after solids contents exceed ~30 vol% solids.

As particle size distribution changes, the porosity defined by a dense pack of its particles changes. The porosity contained in a dense pack defines the volume of carrier fluid required to fill pores. At all solids contents less than the maximum possible, the average distance between particles in suspension varies as the fluid content varies.

The simple way to picture this is to understand that the carrier fluid in a suspension has two functions. It first fills pores. Any fluid in excess of that amount separates particles.

As the particle size distribution changes from batch to batch in a suspension in which the solids content is tightly controlled, the following will occur: the dense-pack porosity will change as the particle size distribution changes; the particles in suspension will be farther apart or closer together depending on the volume of fluid remaining after the pores are filled; and the viscosity of the suspension will decrease or increase as the average distance between particles increases or decreases, respectively.

The primary phenomena that control rheological behaviors in high solids content suspensions are the interactions between particles. When PSDs **cannot** pack well, dense-packed porosities will be high, most of the fluid will fill pores, and the volume of fluid remaining (the non-pore fluid) will be low. In such cases, the InterParticle Separation (IPS) distances will be small, particle-particle interactions will be many and possibly severe, and apparent viscosities will be relatively high.

When PSDs **can** pack well, dense-packed porosities will be small, and only a small volume of fluid will be tied up in the interparticle pores. Non-pore fluid volumes will be relatively high; IPS distances will be large; particle-particle interactions and intensities will be reduced; and apparent viscosities will be relatively low.

The time-average number of collisions and the severity of those collisions are major contributors to the definition of rheological and viscous properties in suspensions.

As batch particle size distributions change from day to day, suspensions may quickly revert from those that flow well to those that flow poorly, and vice versa.

Surface Properties

The nature of the powders and their surface properties also affect the rheological properties of suspensions. This is especially the case when solids contents are sufficiently high that particles collide frequently during suspension flow.

When surfaces are relatively smooth, particles can slide easily against one another. Organic/polymeric additive coatings can reduce interparticle friction and reduce the magnitudes of sliding interactions. When particle surfaces are rough and tortuous, particle/particle friction during collisions can hinder flow and markedly increase suspension viscosities.

Gelation and Flocculation States

Particles in fluids typically exhibit electrostatic surface potentials that cause the particles to repel one another. As electrostatic surface potentials approach zero, Van der Waals forces cause particles to be attracted to one another. In suspensions, interparticle attractive forces can then cause particles to flocculate.

The initial stages of flocculation produce flocs that resemble dust balls on a hardwood floor. As these flocs continue to grow (and they do), they form 3-dimensional networks of particles that span the whole volume of the beaker, container, or tank containing the suspension. Such 3-D structures within suspensions are the characteristic structures produced by gelation.

The level of attractive or repulsive forces between particles in suspension depends on the mineral type of the particles, the pH of the suspension, and the amounts and types of additive chemicals that have been used to alter particle surface properties.

Gelled suspensions which contain these large 3-D structures exhibit yield stresses that must be exceeded before they will flow. If gel structures are strong enough, large particles incorporated into the structures will be prevented from settling. If the gel structures formed are too strong, those suspensions can be extremely difficult (impossible??) to re-fluidize.

Suspensions that exhibit gelation phenomena are usually characterized by shear-thinning rheologies. Gelation phenomena build structures at the same time as shear and flow act to break down the structures. At each different set of shear conditions, suspension viscosities will move towards a dynamic equilibrium point in which the gelation buildup rate balances the shear breakdown rate. Given sufficient time, equilibrium can be achieved and viscosities will then hold relatively constant.

In shear-thinning suspensions at high shear rates, large portions of the gel structure will be destroyed; the sizes of independent flocs of particles will be relatively small; those flocs will move with ease; and apparent viscosities will be relatively low. At low shear rates, less gel structure will be destroyed; independent flocs of particles will be relatively large; the flocs will not be as free to move around; more stress will be required to maintain flow; and apparent viscosities will be relatively high.

Gel structures and flocculation behavior do not apply to simple fluids.

Particle/Particle Collisions and Process Shear Rates

Particle/particle collisions dominate rheological properties in high solids content suspensions, as well as in low solids content suspensions exposed to high shear conditions.

Because the intensities of particle/particle collisions and apparent viscosities of suspensions change as shear rates change, process shear conditions must be considered when pumping, mixing, storing, and handling suspensions. The higher the solids contents of suspensions, the more important it is that such suspensions not be subjected to extremely high shear rates.

This will be discussed in more detail in later chapters. For this discussion, suffice it to say that extremely high shear rates and high solids contents are favorable conditions for producing *dilatancy*. *Dilatant* rheologies are shear-thickening – that is, as shear rates increase, apparent viscosities increase.

Under such conditions particles collide frequently with one another, and viscosities increase quickly as collisions dominate and collision intensities increase.

Generally speaking, dilatant rheologies are unwanted within ceramic process systems because of the many problems they cause.

Dilatant Blockages

In extreme cases, particle/particle collisions can cause *dilatant blockages* to form in which all particles are locked in position and all flow ceases. When a dilatant blockage occurs, more pressure and higher stresses will only further compact and strengthen the blockage. Relaxing the stress on such a blockage, however, does **not** guarantee that it will break up and disappear. In most cases, once a blockage has formed, it won't break up or allow the particles to redisperse.

One of two phenomena usually occurs when a dilatant blockage has formed: (1) Pressure builds up and the blockage is pushed forward, accompanied by severe abrasion where its outer edges drag against the walls and abrade the flow channel; or (2) all flow ceases, and pressures and stresses build up until something breaks. These phenomena will be discussed in more detail in later chapters.

No such problems, however, are possible in simple fluids. These phenomena are only possible when particles are mixed with fluids to produce suspensions.

Unwanted Powder Compaction

To measure the pressure of a compressed gas or a simple liquid, one uses a pressure gauge. Common pressure gauges use Bourdon tubes – C-shaped flexible metal tubes linked mechanically to the gauges' needles. When pressure rises, the curved tubes straighten slightly. This motion, linked mechanically to the needle, corresponds to the pressure.

When a pressure gauge is connected directly into a suspension piping system, it will quickly be ruined because the particles in suspension will be compacted into the flexible Bourdon tube and quickly render it non-flexible. With time, such gauges will become permanently locked in position; they will no longer give any indications of pressure; and they will be ruined.

Special pipe components are available for measuring suspension pressures. Such components contain oil-backed rubber sleeves inside a length of pipe that separate the suspensions within the pipe from the oil within the rubber bladder and the pressure gauge. Pressure gauges in these components, which are filled with oil, sense the oil pressure. As the suspension flows through the pipe, the pressure is transmitted through the rubber bladder to the oil, so the pressure of the oil is equal to the pressure of the suspension.

Special components like these, which are designed to work with suspensions, are necessary additions to suspension piping systems. They are considerably more expensive than the pipe fitting and pressure gauge that we are accustomed to using on simple fluids, but they will not be ruined by the particles in particle/fluid suspensions.

Abrasion

Another problem that occurs when suspensions are pumped is that particles abrade pipes and anything in them. For example, the author's research team once tested both an in-line viscometer and an in-line high intensity dispersion head in the piping system of a high solids suspension in a pilot production plant. Both viscometer and mixing heads were ruined by abrasion within an hour of their installation and initial use.

Pumps used with suspensions are also known to quickly be abraded by particles. Gaps between impellers and pump bodies, which are supposed to be relatively small, increase quickly in size due to abrasion. As they do, pump output pressures decrease. When suspensions come in contact with pump and mixer bearings, the bearings tend to wear out quickly — sometimes, as the author has experienced, within minutes.

Many components that work well in the piping systems, pumps, and mixers of simple fluid systems are not sufficiently robust to withstand the abrasive environments in suspension systems.

Suspension Stability

Settling

Simple fluids don't contain anything that can settle, so stability is not a consideration. Particles in suspension, however, can and do settle, so the stability of suspensions must be taken into account when processes are being designed.

In process storage tanks, suspensions require agitation to produce recirculation (particularly upward flow in the tank) to counter settling tendencies. Some agitators produce upward tank flow. Other agitators do not[7].

As an example, consider a relatively tall, small diameter, process tank that was used to volumetrically dispense suspension in a batch operation. This tank was not properly agitated. Over a period of time, the specific gravity in the bottom of the tank increased significantly above the average and the specific gravity at the top of the tank decreased significantly below the average. Suspension was dispensed from the bottom of this tank so the specific gravity of suspension varied from high to low gravity as the tank was emptied and as portions of the tank were fed to the batch operation. Until slurry stability was considered in this process, and changes were made, constant volume additions from this tank were not supplying constant particle mass to the main production batch.

Flow Velocities in Piping Systems

Another phenomenon that must be considered in piping systems is the actual flow velocity in a pipe. High viscosity fluids like molasses can be pumped through relatively large diameter pipes without any settling problems.

Particles in suspension, however, can settle in a pipe if the flow velocity is too low (or zero). Over time, such settling will decrease the effective diameter of the pipe, and it could lead to eventual blockage of the pipe. Depending on the flow rate and suspension properties, large particles can settle out while the fines are delivered to the process. When settling occurs in a pipe as a function of particle size, the particle size distribution (PSD) of suspension dispensed from the piping system will not be identical to the PSD of suspension fed into the pipe.

As an example of this, the author has seen a suspension piping system designed with a single stainless steel pipe from the storage tank to the process site. Flow velocities in this pipe, which included a fairly long, horizontal run, were extremely low. The process was run intermittently, so there were large periods of time with no flow at all. During these times, particles in the suspension settled in the pipe. If this pipe had contained a simple fluid instead of a suspension, the design would have worked well.

Figure 3.2A shows a diagram of this system as designed and installed. Figure 3.2B shows how this design should have modified to accommodate the suspension. Figure 3.2A shows the tank, pump, and the long horizontal run to the process. A valve at the process controlled flow to the process. This is a fine design for simple fluids.

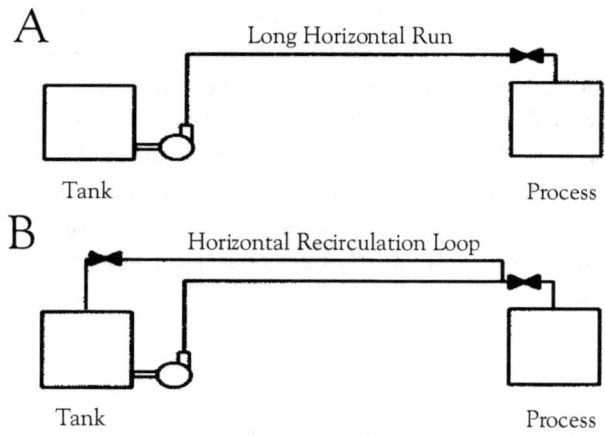

Figure 3.2 (A) Piping System for Simple Fluids
and (B) Showing Changes to Accommodate Suspensions

The pipe from the pump to the process was a relatively large diameter (1") stainless steel pipe. Had this been a low viscosity simple fluid (water, for example), a 1/4" copper tube would have been sufficient.

These two features, the larger diameter and the stainless steel pipe, may have been the designer's accommodations to convert the simple fluid design to one that could handle a suspension. The stainless steel pipe was a good choice. The increased pipe size and the long, dead-ended line from tank to process were not appropriate for most suspension systems.

To accommodate the suspension, the design should have been like the one shown in Figure 3.2B. Stainless steel pipe is a good choice, but the pipe diameter could be smaller (than 1" diameter) to maintain adequate flow velocities to prevent particles from settling. The recirculation loop to the process and back to the tank allows suspension flow to be continuous. If the suspension flow velocity is adequate to prevent particle settling, continuous flow won't allow particles to settle in the main pipe.

The only place particles can settle in the system shown in B is between the process valve and the process. When this distance is as short as possible, the possibility settling problems will occur will have been minimized.

The piping system in Figure 3.2A will work fine for a high viscosity simple fluid, but it is not fine for a high viscosity suspension.

No extraordinary features, such as the continuously running recirculation loop shown in Figure 3.2B, are necessary when a piping system contains only a simple fluid.

Fluid Effects on Rheology

There is not much that can be done to the carrier fluids in suspensions to change the rheological behaviors of those suspensions. It is always possible, however, to add more fluid to dilute a suspension that is causing problems.

Many ceramists consider their particular water content (that is, the specific gravity of their process suspension) to be *sacred*. It can't be touched or changed. In the author's opinion, this should be carefully rethought. If one is going to hold a suspension characteristic constant, it appears that it should be the InterParticle Spacing (IPS — i.e., the average distance between particles in the suspension) that is held constant, not the fluid content.

As a process particle size distribution (PSD) changes from batch to batch, IPS will vary when the solids content remains constant. As the PSD changes from batch to batch, the IPS *can* remain constant if the solids content is varied in response to the changes in packing capability of the PSD.

The IPS is important if one's goal is to maintain similarity of interactions between particles in suspension. If IPS remains constant, the average distance between particles over which electrostatic and other interparticle forces act, remains constant. Under such conditions, particle/particle interactions and interparticle attractive/repulsive phenomena can remain similar from batch to batch.

When IPS is large one day and small the next, corresponding to low viscosity one day and high viscosity the next, the tendency is for ceramists to adjust process viscosities by adjusting additive chemical concentrations. Adjustments to additive flocculant and defloculant concentrations *are* used to control the rheological properties of ceramic suspensions and bodies. In this example, however, using chemical additives to correct for IPS problems (which is a particle physics problem, not a chemical problem) is an attempt to implement a chemical solution to a particle physics problem. Sometimes such solutions will work. Sometimes they will not.

Such solutions actually address symptoms rather than the problems themselves. If the effective viscosity of a suspension increases because the particles are closer today than they were yesterday (IPS is smaller), changing the surface properties of the particles with deflocculants may help to reduce the effective viscosity (the symptom), but it does nothing to increase the IPS (the underlying problem).

At the extremes, such chemical adjustments can cause individual process suspensions at constant apparent viscosity at a particular shear rate to range from highly flocculated one day to highly deflocculated the next. Highly flocculated suspensions tend to be shear-thinning, whereas highly deflocculated suspensions tend to be dilatant. Daily chemical adjustments, as shown in this example, can cause major day to day swings in suspension rheology and in process performance.

To recognize that the PSD has changed, and to make corrections to it, requires that the complete PSD be precisely and routinely monitored. Many companies do not sufficiently monitor, nor attempt to control, PSD. Making adjustments to correct fluctuating PSDs is not easy, but such corrections would address problems rather than symptoms.

When PSD adjustments are not an option, dilution still remains a possibility. As the IPS in process suspensions decreases and viscosities increase, dilution can increase the IPS without making any changes to PSD or to chemical additive concentrations.

Once again, none of these considerations are necessary when dealing with simple fluids.

Chemical Additive Effects on Rheology

Chemical additives are routinely used to *deflocculate* suspensions to reduce their viscosities, or to *flocculate* suspensions to increase their viscosities. Deflocculated suspensions have very different characteristic rheological properties than flocculated suspensions. Deflocculated suspensions also have very different behaviors in process environments than flocculated suspensions. These characteristics and behaviors should be well understood when chemical additives are used in ceramic suspensions and bodies.

To show the effects of chemical additives, the discussion will focus on extreme cases, recognizing that most ceramic process bodies exhibit more moderate properties between the extremes.

Flocculation and Deflocculation

Just to make sure we're all on the same playing field, the terms *flocculation* and *deflocculation* need to be defined. Professor Funk's definitions[8] are most appropriate for this. When shepherds watched their flocks of sheep by night, the sheep would *flock-ulate* as they all gathered closely around the shepherd. If the shepherd were not present, and a wolf jumped into the middle of the flock, the sheep would *de-flock-ulate* and scatter.

Flocculated particles (note the proper spelling) are attracted to one another. They will approach one another closely, first forming small flocs and eventually forming larger 3-D structures.

Deflocculated particles (note the proper spelling here, too) repel one another. They scatter and take positions as far from each other as possible. Even when fluid contents decrease and particles are forced closer to one another, deflocculated particles will remain as far apart as possible within the confines of the new particle physics properties.

Particles in *partially* flocculated or *partially* deflocculated systems attract or repel one another, but with less than the maximum possible attractive or repulsive forces.

Extreme Deflocculation

Within suspensions that are highly deflocculated, all particles will strongly repel one another. This happens when their surfaces are electrostatically highly charged either positive or negative (absolute values of zeta potentials are $> \sim 60mV$). Like-charges repel. So under such conditions, all particles will repel and remain as far apart as possible.

In highly deflocculated suspensions where most particles travel independently of one another, fine particles preferentially travel with the fluids. The fines tend to be dragged along with the fluid flow. Coarse particles which have relatively large masses and momenta, will be relatively unaffected by the motion of the carrier fluid. As a result, the powders in such suspensions will segregate (un-mix) by size during flow. Gelation will be minimal and the cushioning effects of gel structures will not be present. Without the cushioning effects of gelled structures, particle/particle collision energies during flow will be relatively high, and the suspensions will tend to be dilatant, even under low shear conditions.

Dilatant suspensions are prone to form dilatant blockages, as discussed earlier. When dealing with dilatancy, and especially when trying to prevent such blockages from forming, shear conditions in highly deflocculated suspensions should remain low.

When highly deflocculated suspensions are dewatered, the finest particles which travel freely with the fluids tend to flow to the filter interface where they are deposited. If initial dewatering rates are high, filter cakes can form in layers with the finest particles at the filter surface, followed by layers of the next larger particles. The size of the pores in the first layer of predominantly colloidal particles will be very, very small. The porosity in this layer will be high, but the pore sizes will be very small.

If colloidal particles form the first layer deposited in a filter pressing or slip casting operation, the overall dewatering rate will be slow because all fluid to be removed must pass through that first layer. Filter pressing, slip casting, drying, and any other dewatering operation, in which the water must pass through a layer with such small pores, will be extremely slow.

As wares are formed from highly deflocculated slurries, particles will remain mobile as long as possible. Structures between such particles will be slow to form. When the particles finally do contact one another, pores throughout the structures will usually be very small.

Compacts formed from deflocculated bodies tend to be dense; they will dry slowly because the pores are small; and they will exhibit relatively low firing shrinkages.

Strong Flocculation

When suspensions are highly flocculated, the particles' electrostatic surface charges will be near zero. This usually occurs when multivalent cations such as Ca^{+2}, Mg^{+2}, or Al^{+3} in the form of soluble and partially-soluble salts of Cl^-, SO_4^{-2}, and CO_3^{-2}, for example, are added to the suspensions. As electrostatic surface charges decrease towards zero and repulsive forces disappear, Van der Waals forces of attraction take over and pull the particles together to form gel structures.

Highly flocculated suspensions can be expected to exhibit strong gelation behavior. Some inorganic salts are only partially soluble in water, so the cations may be released into suspensions over long periods of time (hours to days).

Gel structures in highly flocculated suspensions build quickly and generally break down easily during flow. Shear-thinning behavior is typical of flocculated suspensions. During flow, particles in flocculated suspensions usually do not exhibit any tendency to segregate by size. The attractive gelation phenomena pull all particles together into the forming gel structure. These attractive forces are especially strong on the smallest particles. So unlike the behavior in deflocculated suspensions, colloidal particles in flocculated suspensions do not flow with the fluid but they are quickly tied up and immobilized by the gel structure.

The interparticle porosity in flocculated systems is characterized by relatively large, open pore channels. This makes it relatively easy for fluids to flow through gel structures. Filter pressing, casting, and drying operations proceed easily and quickly in flocculated suspensions and forming bodies.

Large yield stresses are also characteristic of highly flocculated suspensions. When suspensions are allowed to sit quiescently for long periods of time, strong gel structures can build that are difficult to overcome when the suspensions are again to be pumped or otherwise processed.

Extreme Flocculation and Syneresis

Syneresis can occur in suspensions when flocculation levels are extreme. When syneresis occurs, particles are pulled tightly together into relatively dense gel structures from which interparticle fluids are expelled.

In *syneretic* suspensions, the expelled fluids are usually clear supernatant, and as gel structures densify, relatively large cracks are produced in the gel structure as fluid continues to be expelled. When syneresis occurs in forming bodies, extrusion blanks and filter press cakes can crack and break into pieces. They can literally fall apart.

When syneresis is visible in extrusions and filter cakes, it will continue to cause problems after the wares have been formed.

Desired Flocculation/Deflocculation States

It is **never** desirable to have either of the extremes – extreme deflocculation or extreme flocculation – in ceramic process systems. The desired states for ceramic forming bodies and slips should be **partially** flocculated or **partially** deflocculated states. The key word here is *partially*.

Making adjustments to additive chemistries to achieve a particular process *viscosity* at a known shear rate, without paying attention to the body's *rheological properties*, is dangerous. PSD and other particle physics characteristics can change sufficiently from batch to batch to cause wide swings in the state of deflocculation/flocculation after chemical adjustments are made. It may be possible to achieve the control viscosities by adjusting the levels of flocculant or deflocculant additions, but the resulting rheologies can vary widely and the bodies may exhibit wild processing behavioral swings as a result.

Suspension *viscosities* and *rheologies* should both be monitored to guarantee the similarity of process slip properties from batch to batch, and from day to day.

Summary

In this chapter, the discussion focused on the differences between simple fluid properties and suspension properties. Many considerations that must be applied to suspension processing systems, such as those discussed in this chapter, are totally foreign to the scope of considerations that apply to simple fluids.

As we continue through this text, more details will be provided concerning the phenomena discussed in this chapter that apply to suspensions, but have no application to simple fluids.

Chapter Four

Time-Independent Rheologies

There are two general categories of rheology: time-independent and time-dependent. In this chapter, we will introduce and discuss the time-independent rheologies.

Figure 4.1 shows six time-independent rheologies. They are considered to be independent of time because the duration of shear has no effect on these rheological properties. Dilatant (shear-thickening) behaviors are characterized by increasing apparent viscosities as shear rates increase. Pseudoplastic (shear-thinning)

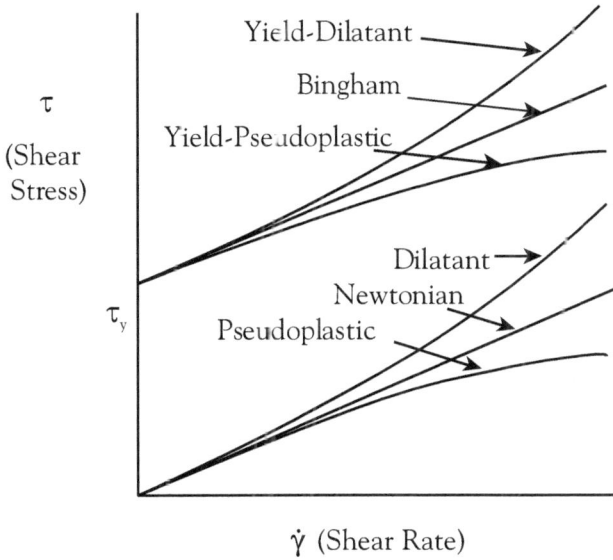

Figure 4.1 Time-Independent Rheologies

behaviors are characterized by decreasing apparent viscosities as shear rates increase. The ratios of shear stress to shear rate remain constant for Newtonian fluids, and their rheograms exhibit linear behavior beginning at the origin.

The same three types of behavior also occur after the applied stress has exceeded a yield stress value. These three are known as yield-dilatant, yield-pseudoplastic, and Bingham rheologies.

Five of the six rheologies shown in Figure 4.1 are non-Newtonian rheologies. The sixth rheology is the simple, constant-viscosity, Newtonian rheology. All five rheologies in which the ratios of shear stress to shear rate are not constant over the whole range of shear rates are known as non-Newtonian fluids.

Apparent Viscosity

The *apparent viscosity*, μ_a, of a suspension is defined as the ratio of the measured shear stress to the applied shear rate, and it applies specifically at the measurement conditions. One can then say, for example, that a suspension has an apparent viscosity of 1000 mPa · s at 250 s^{-1}. Note that the viscosity, and the shear conditions under which it was measured, are both given. The apparent viscosity and the shear rate at which it was measured are both required to describe any non-Newtonian fluid.

Figure 4.1 shows a single Newtonian fluid. All other Newtonian fluids will also have linear behaviors starting at the origin (zero shear stress/zero shear rate). The linear behaviors of other Newtonian fluids will be characterized by different slopes. Refer again to Figure 2.3. High viscosity fluids will have steeper slopes than the fluid shown in Figure 4.1. Shear stresses will increase quickly at relatively low shear rates. Low viscosity fluids will have lower slopes than the fluid shown in Figure 4.1. The rheograms of low viscosity fluids will remain relatively close to the shear rate axis, achieving high shear rates with small applied shear stresses.

Apparent viscosities of non-Newtonian fluids have similar characteristics. All points on a rheogram that are in the upper left region of the diagram where shear stresses are high, but shear rates are low, will exhibit relatively high apparent viscosities. All points on a rheogram that are in the lower right region of the diagram where shear stresses are low and shear rates are high will exhibit relatively low apparent viscosities.

Note that the *viscosity*, or more precisely the *apparent viscosity*, of a fluid or suspension is **not** the slope of the rheogram at any point, but the ratio of the shear stress to the shear rate at any point. It happens that the shear stress to shear rate ratio and the slope of the rheogram are identical in Newtonian fluids, but the two are usually very different in non-Newtonian fluids. The definition of viscosity to remember is that the apparent viscosity of any fluid is **the ratio of the shear stress to the shear rate** at the point of measurement on the rheogram.

Yield Stress

Three non-Newtonian fluid categories in Figure 4.1 exhibit yield stresses. In each case, the applied stress must exceed the yield stress before flow can begin. Simple fluids do not exhibit yield stresses because, by definition, a simple Newtonian fluid will flow under any applied stress, regardless how small.

Ceramic bodies must all have yield stresses if wares are to hold their shapes after they are formed. Without yield stresses, ceramic wares could not be formed and there would be no ceramic industry as we know it. Some bodies, ware shapes, and certain forming processes require stronger yield stresses than others, but **all** ceramic production bodies will exhibit yield stresses.

Pseudoplastic (Shear-Thinning) Fluids

Apparent viscosities of pseudoplastic fluids decrease as shear rates increase. Note that the issue is not whether the rheogram starts at zero shear rate with a higher or lower apparent viscosity than the Newtonian fluid shown in Figure 4.1. The issue is which way the rheograms curve as shear rates increase. For shear-thinning suspensions, the rheogram will curve towards the shear rate axis.

When rheograms are plotted as in Figure 4.1 showing shear stress versus shear rate, the rheograms of pseudoplastic fluids will curve towards the horizontal as shear rates increase. In pseudoplastic fluids and suspensions, less and less incremental addition of shear stress is required to achieve higher and higher shear rates.

Dilatant (Shear-Thickening) Fluids

Apparent viscosities of dilatant fluids increase as shear rates increase. In plots of shear stress versus shear rate as in Figure 4.1, the rheograms of dilatant fluids will curve towards the vertical as shear rates increase.

In dilatant fluids, more and more incremental addition of shear stress is required to achieve higher and higher shear rates. The extreme case for dilatant fluids occurs when so much shear stress is applied that all shear and flow stop. This is known as a dilatant blockage.

Dilatant blockages were mentioned briefly in Chapter 3. Just before a dilatant blockage forms, shear stresses will increase quickly and the rheogram trace will take off towards the vertical. At the instant the blockage forms and shear and flow stop, the rheogram no longer applies. The reason is that many viscometers will continue to indicate relatively high shear conditions, even when particles are locked together in a blockage in which no shear is taking place.

Depending on the design of the viscometer being used, a dilatant blockage may slide against the viscometer surfaces which will produce false readings. Viscometers, if not designed properly, may be damaged when dilatant blockages occur.

Bingham Suspensions

The rheogram labeled *Bingham* in Figure 4.1 corresponds mathematically to a Newtonian fluid that exhibits a yield stress. The Bingham rheogram is the ideal rheogram of a *plastic* material. It is questionable, however, whether ideal Bingham fluids actually exist. Many suspensions have been labeled Bingham that are more likely yield-pseudoplastic or yield-dilatant.

The most attractive feature of Bingham fluids is the simplicity of their equations. A Newtonian rheogram exhibits linear behavior through the origin, as shown in Equations (2-9) and (2-12). The Bingham equation is identical, but with the addition of the yield stress as shown in Equation (4-1):

$$\tau_s = \mu \, \dot{\gamma} \tag{2-12}$$

$$\tau_s = \mu_B \, \dot{\gamma} + \tau_y \tag{4-1}$$

$$Y = mX + b \qquad (4\text{-}2)$$

Equation (4-1) is the equation of a straight line with the slope equal to the viscosity, μ_B, and the Y-intercept equal to the yield stress. In this equation, the slope of the line, μ_B, is called the Bingham viscosity.

Mathematically, the Bingham equation is a Newtonian fluid with a shift of origin so the rheogram intercepts the shear stress axis at the yield stress value. The Bingham viscosity in Equation (4-1) is the slope of its straight line, as the Newtonian viscosity is the slope of its straight line in Equation (2-12). The Bingham rheology is popular because of this mathematical simplicity.

Measured apparent viscosities of Bingham suspensions, however, are very different from the Bingham viscosity, μ_B, in Equation (4-1). This can be quite confusing. Measured apparent viscosities are always the ratios of applied shear stresses to the shear rates achieved. The Bingham viscosity is the mathematical slope of the Bingham equation that characterizes the measured data.

Apparent viscosities should never be confused with Bingham viscosities. They are calculated differently and they have different physical meanings. Apparent viscosities are useful in the plant and lab. Apparent viscosities only approach the value of the Bingham viscosity at the high shear rate limit of the apparent viscosity. So the Bingham viscosity is only meaningful in a practical way at high shear rates. At most process shear rates, Bingham viscosities will be considerably lower values than measured apparent viscosities.

The simplicity of the Bingham equation allows it to be used, with some simplifying assumptions, to calculate yield stresses of measured yield-rheograms over the shear rate ranges of interest. To do so, two measured shear stresses (at two shear rates) can be used to define the straight line Bingham rheogram which can then be used to calculate the yield stress for that fluid. We used such procedures in our textbook[9] to calculate yield stresses as well as gelation and other rheological parameters for ceramic suspensions.

Viscosity versus Shear Rate

Figure 4.1 shows the six rheologies plotted as shear stress versus shear rate. Figure 4.2 shows these same six rheologies plotted as apparent viscosity versus shear rate.

The yield stress of Figure 4.1 corresponds to the extremely high starting viscosities as shear rates approach zero in Figure 4.2. From these initial high viscosities, the viscosities of all three yield-rheologies decrease quickly and approach their counterpart non-yield-stress rheologies as shear rates increase.

The Newtonian rheology on Figure 4.2 is a horizontal line that indicates constant viscosity over the whole range of shear rates. Apparent viscosities of Bingham suspensions approach the Newtonian constant viscosity behavior at high shear rates. Yield-dilatant apparent viscosities approach dilatant viscosity behaviors at high shear rates. Yield-pseudoplastic apparent viscosities approach pseudoplastic viscosity behaviors at high shear rates.

Figures 4.1 and 4.2 are two different ways to plot the same six time-independent rheograms. Either form is acceptable.

Making Time-Independent Measurements

It is difficult to measure time-independent (TI) rheologies without influences from time-dependent (TD) phenomena. *Time-dependent* behavior

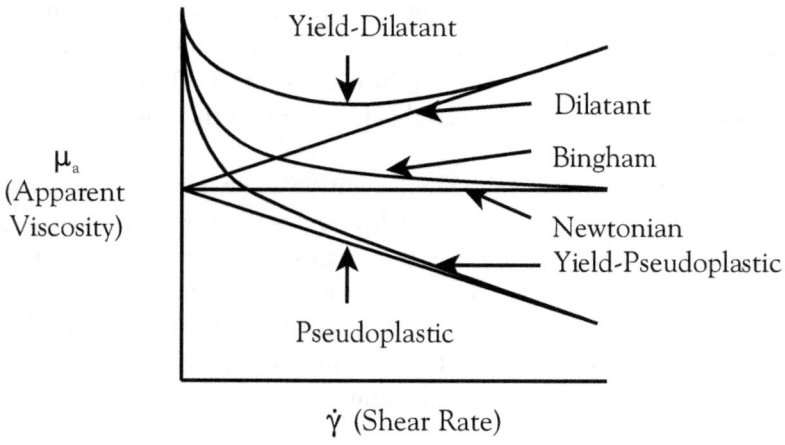

Figure 4.2 Time-Independent Rheologies Plotted as
Apparent Viscosity versus Shear Rate

is affected by shear histories. That is, the magnitude and time duration of applied shear affect the measured apparent viscosities. It is difficult to make apparent viscosity measurements instantaneously, and it's impossible to make them at all without imposing a shear history on suspensions during the time it takes to make the measurements. The mere act of making a measurement imposes a shear history on the suspension.

To measure TI rheologies (without TD effects), one needs to make measurements instantaneously (or as close to this as is practically possible) without imposing any shear history on the suspension being measured. This is difficult to successfully accomplish.

When an apparent viscosity of a suspension is measured on a viscometer, some shear must be applied for a period of time. If the suspension exhibits TD properties, the measured apparent viscosities will be affected by the shear history (regardless how brief) imparted by the viscometer. Shear*time history is a TD effect which will be discussed in the next chapter.

To isolate and accurately measure TI rheological properties of a suspension, several representative samples of that suspension must be prepared to make the measurements. Sample preparation procedures and shear histories for each sample prior to taking measurements should be identical. Each sample should then be measured at a different single shear rate and each measurement should be performed in as short a measurement time as possible.

The results from these several samples, each measurement of which corresponds to a single shear rate, can then be combined to

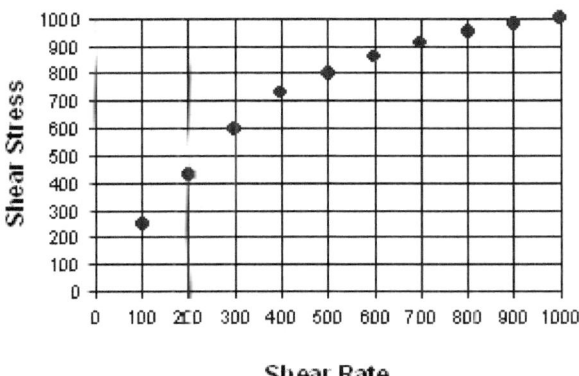

Figure 4.3   Shear-Thinning Time-Independent Rheogram
Consisting of Single Measurements on Ten Samples

form a rheogram that shows the TI behavior of the sample over the range of measured shear rates. Figure 4-3 shows a sample rheogram of a shear-thinning, thixotropic suspension. Note that each point on the rheogram **must** be measured on a single, independent, but identically prepared sample.

Attempts to measure TI behavior over a range of shear rates (sometimes called a 'shear rate program') in one long measurement will show the combined effects of TI and TD rheologies. Figure 4-4 shows an example of such a rheogram. If a viscometer is set up to measure apparent viscosities as

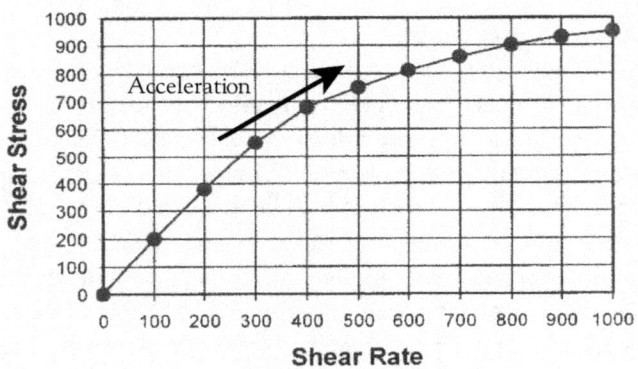

Figure 4.4   Shear-Thinning Rheogram Measured in a 10 Minute Acceleration from 0 to 1000s^{-1}

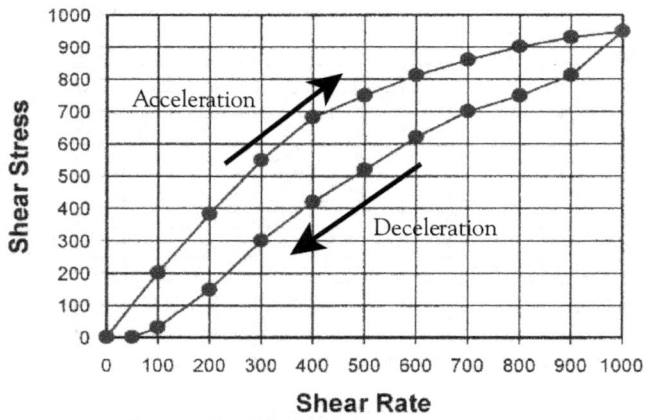

Figure 4.5   Shear-Thinning Rheogram Measured in a 20 Minute Acceleration/Deceleration from 0 to 1000 to 0s^{-1} Showing Time-Independent and Time-Dependent Behavior

the shear rate accelerates from 0 to $1000s^{-1}$ in a period of 10 minutes, the 10 minutes of shear history will affect the measured viscosities at all shear rates. Note that all measured shear stresses in Figure 4.4 are lower than their counterparts in Figure 4.3. This is due to the shear histories imposed during the long continuous measurement in Figure 4.4 that was not present when individual points are measured as in Figure 4.3.

Figure 4.5 shows what can happen when this measurement immediately continues for another 10 minutes as shear rate decelerates again from 1000 to $0s^{-1}$. Time-dependent thixotropic effects can define a new rheogram in which all measured viscosities during the deceleration are at lower viscosities than their acceleration counterparts.

Under 10 minutes of constant acceleration followed by 10 more minutes of constant deceleration during a 20 minute measurement, each increase or decrease of $100s^{-1}$ takes a minute. During the 20 minute program, it takes 1 minute to accelerate to a shear rate of $100s^{-1}$ and it takes 18 more minutes to accelerate to $1000s^{-1}$ and then decelerate again to $100s^{-1}$.

The 18 additional minutes of shear history that occurred between the two $100s^{-1}$ measurements, as shown in Figure 4.5 will produce a lower measured viscosity at the second, decelerating measurement when the shear-thinning (TI) fluid is also thixotropic (TD).

When the acceleration and deceleration rheograms differ, as shown in Figure 4.5, this is known as *hysteresis*. Hysteresis in a measured rheogram is caused by the presence of time-dependent effects. When the suspension rheology is not time-dependent the deceleration measurement will retrace the acceleration measurement and will not show any hysteresis. Most non-Newtonian particle/fluid suspensions, however, usually exhibit both time-independent and time-dependent character.

If another sample of the suspension shown in Figure 4.5 were measured from 0 to $1000s^{-1}$ in a period of 5 minutes, measured apparent viscosities would differ from those in the 10 minute acceleration rheogram. The time-independent behavior would be the same, but the time-dependent behavior would differ because the acceleration was doubled, and the time duration of the measurement was cut in half. Measured viscosities would be slightly higher in a 5 minute measurement than in the 10 minute measurement because the imposed shear history in a 5 minute run is less than that in a 10 minute run.

Summary

Time-independent rheologies show the changes in viscosity that derive directly from changes in shear rate. For ceramists, the important time-independent rheologies are the yield rheologies. The yield stress, which is characteristic of these three rheologies, allows ceramic wares to hold their shapes after forming.

Time-dependent effects cannot easily be separated from time-independent effects during rheological measurements. To measure time-independent effects unhindered by time-dependencies, several samples of a single suspension must each be measured at single (but different) shear rates, and the several results must then be combined to form a time-independent rheogram.

Chapter Five

Time-Dependent Rheologies

There are two time-dependent (TD) non-Newtonian rheologies that characterize suspensions and fluids: *thixotropy* and *rheopexy*. Examples of these two rheologies are shown in Figure 5.1.

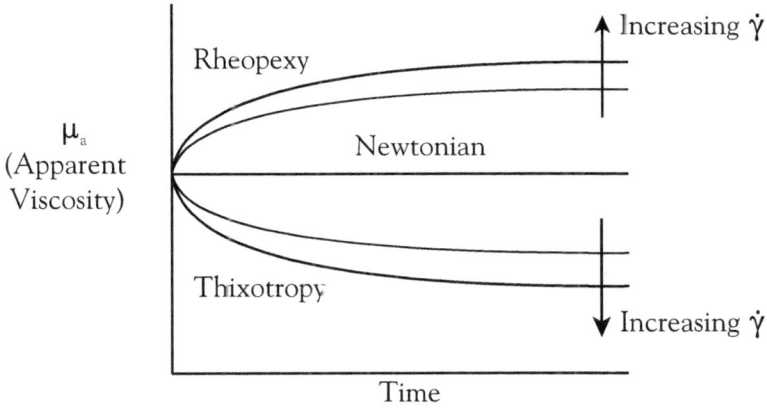

Figure 5.1 Time-Dependent Rheologies
Measured at Constant Shear Rate

Thixotropy

Apparent viscosities of *thixotropic* suspensions decrease and approach minimum, limiting viscosities as the suspensions are exposed to constant shear rates with time. The thixotropic rheograms are the two lower curves shown in Figure 5.1. As intensities of shear conditions increase, the measured, limiting apparent viscosities of thixotropic suspensions decrease. Beyond some upper shear rate, however, further increases in shear conditions no longer produce corresponding decreases in the limiting viscosity.

Rheopexy

The apparent viscosities of *rheopectic* suspensions increase with time under constant shear conditions. The rheopectic rheograms are the two upper curves in Figure 5.1. Higher shear rates produce increasingly viscous behavior with time, up to the limit of no flow at all.

Shear History

The apparent viscosities of time-dependent fluids and suspensions respond to the length of time of exposure and the rate of shear. The shear history is the imposed shear rate times the time of exposure. It can also be described as the area under a shear rate versus time curve. As such, it is a dimensionless number:

$$N_{SH} = \text{shear rate} \cdot \text{time} [=] \ s^{-1} \cdot s = \text{dimensionless} \qquad (5\text{-}1)$$

The greater the shear history, the more time the intensity of shear has to work on the structure of a suspension and raise or lower its apparent viscosity.

Consider again the sample rheogram shown in Figure 4-5. That figure depicts a rheogram measured as the suspension was exposed to constant acceleration from 0 to $1000s^{-1}$ in 10 minutes followed immediately by constant deceleration to $0s^{-1}$ over the next 10 minutes. Figure 5.2 shows the shear rate versus time program for that run.

When apparent viscosity is measured in a single acceleration/deceleration program as this, the shear history after the full 20 minute run is exactly double that of the shear history imposed over the first 10 minutes. The area under the curve for each 10 minutes of the run is the area of a triangle. The shear history for the constant acceleration part of the program is:

$$\text{Shear History}_{\text{Acceleration}} = \tfrac{1}{2}bh = \tfrac{1}{2}(600s)\,1000s^{-1} = 300{,}000 \qquad (5\text{-}2)$$

The shear history during deceleration is the triangular area for the second half of the program which has the same value as that just calculated in Equation

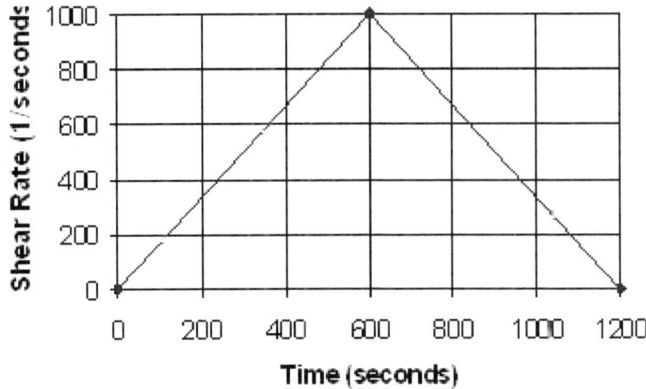

Figure 5.2 A Typical Rheometer Measurement Program:
10 minutes of constant acceleration to a maximum
shear rate followed immediately by 10 minutes of
constant deceleration back to 0 shear rate.

5.2. The total shear history of the program is the sum of these two:

$$N_{SH\ Total} = N_{SH\ Acceleration} + N_{SH\ Deceleration} =$$
$$= 300,000 + 300,000 = 600,000 \qquad (5\text{-}3)$$

This calculation demonstrates that each half of this type of acceleration–deceleration measurement program contributes exactly the same shear history to the fluid or suspension being measured.

Consider, however, the shear histories at the two points in the program at which apparent viscosities are measured at $100s^{-1}$. One is measured after 60 seconds of acceleration and the other is measured after 1140 seconds (600 seconds of acceleration plus 540 seconds of deceleration).

The shear history after one minute of acceleration when the shear rate reaches $100s^{-1}$ is:

$$N_{SH\ 1} @ 100s^{-1} = \frac{1}{2} (60s)\ 100s^{-1} = 3000 \qquad (5\text{-}4)$$

After 10 minutes of constant acceleration from 0 to $1000s^{-1}$ and 9 more minutes of constant deceleration from $1000s^{-1}$ back to $100s^{-1}$, the shear history again at $100s^{-1}$ is:

$$N_{SH\ 2}\ @\ 100s^{-1} = N_{SH\ Acceleration} + N_{SH\ Deceleration} =$$
$$(\tfrac{1}{2}\ (600s)\ 1000s^{-1}) + (\tfrac{1}{2}\ (600s)\ 1000s^{-1} - \tfrac{1}{2}\ (60s)\ 100s^{-1}) =$$
$$(300{,}000) + (300{,}000 - 3000) = 597{,}000 \qquad\qquad (5\text{-}5)$$

The difference between the $100s^{-1}$ shear history during acceleration (3,000) and the $100s^{-1}$ shear history during deceleration (597,000) is substantial. The shear history after 19 minutes when the shear rate decelerates to $100s^{-1}$ is 199 times the amount to which the suspension was exposed during the first minute of acceleration to $100s^{-1}$.

Time-dependent suspensions, which respond to the shear*time history, will exhibit very different apparent viscosities after such different shear histories, even when the apparent viscosities are both measured at the same shear rate.

Shear history is not usually quantified in this way, but these sample calculations show that shear histories can be substantially different, even within relatively common viscometer measurement programs.

Gelation and Thixotropy

Thixotropic rheologies are typical of flocculated, shear-thinning suspensions which exhibit gelation behavior. When shear is applied to a thixotropic suspension, gel structures break down and apparent viscosities decrease. When shear is discontinued and the suspension is allowed to sit undisturbed, gelation phenomena rebuild structures throughout the suspension and apparent viscosities increase again.

When a gelled suspension is sheared, a dynamic balance occurs between the gelation which builds structure and increases viscosity, and the shear conditions which break the gel structure and decrease viscosity. Gelation proceeds at a rate that is controlled by the properties of the suspension components (additive type and concentration, flocculation/deflocculation state, interparticle spacing, etc.) Gel breakdown rate is a controlled by the rate of imposed shear.

As a gelling suspension is sheared, the rates of the two phenomena (gel buildup rate and gel breakdown rate) will find a balance point at each shear rate. When the rates of these two phenomena are equal, the TD rheograms will exhibit constant viscosity with time as shown by the relatively constant, limiting viscosities in Figure 5.1.

Higher shear rates will destroy more gel structure and suspensions will then exhibit lower apparent viscosities. When apparent viscosities no longer decrease with further increases in shear rate, this is an indication that all of the gel structure has been destroyed. Under such conditions, particles travel as individuals, rather than as small flocs that are remnants of the gel structure. When shear conditions are reduced, gelation will again cause flocs and 3-D structures to form.

When gelation occurs in a suspension, the gel-building phenomena cannot be **stopped** simply by imposed shear, but gelation can be **overpowered** by high shear conditions. High shear conditions can break up gel structures faster than gelation phenomena can build them. When the intensity of shear decreases, the gelation phenomena will once again become apparent as apparent viscosities rise.

Rheopexy and Particle/Particle Collisions

In the introductory chapter, it was mentioned that corn starch in water is an example of a dilatant suspension. Apparent viscosities increase in dilatant suspensions as shear rates increase. *Dilatancy* is the time-independent (TI) shear-thickening rheology. Its time-dependent (TD) shear-thickening counterpart is *rheopexy*.

Dilatancy and rheopexy in ceramic suspensions are the result of particle/particle collisions. As shear intensities increase, the magnitude of particle/particle interactions and collisions increase, and measured apparent viscosities increase. The TI dilatancy phenomena occurs independent of time as shear rates increase. The TD rheopexy phenomena occurs with time at constant shear rate.

Just as pseudoplasticity and thixotropy are related to gelation behavior, dilatancy and rheopexy are related to particle/particle interactions and collisions during shear.

Anything that increases the intensity of particle/particle interactions during flow can increase dilatant and rheopectic effects. Deflocculated suspensions tend to exhibit dilatancy, and if it were easy to measure, most such suspensions would be found to be rheopectic as well.

Rheopexy is not easy to measure and it is rarely seen in viscometer measurements. Its TD rheograms, as shown in Figure 5.1, exhibit increasing apparent viscosities that build up to maximum levels at each set of applied shear conditions. Higher levels of applied shear produce greater measured apparent viscosities. The ultimate limit occurs when particles lock up into a dilatant structure and shear and flow come to an abrupt halt. Such blockages, known as *dilatant blockages*, will be discussed in more detail in a later chapter.

There are several reasons for the lack of measurements that confirm rheopexy in suspensions. Deflocculated suspensions that could show rheopectic properties tend to be unstable – i.e., particles settle quickly. If such suspensions are treated gently and with deliberation as measurements are being taken, the particles will settle before the measurements are completed.

Suspensions that could exhibit rheopexy will almost certainly also exhibit dilatancy. The process of taking rheological measurements on dilatant and rheopectic suspensions can easily damage viscometers. Blockages can clog narrow passageways, and the onset of such blockages can ruin viscometer measuring heads as well as control motors and gearing that are not sufficiently protected with appropriate clutch mechanisms.

The magnitude of measured apparent viscosities in such suspensions can also quickly exceed the measurement capabilities of viscometers. If the target viscosity for a production body is well within the capabilities of a particular viscometer, the mere fact that measurements are headed out of range and will exceed the viscometer's measurement limits are frequently sufficient to cause process engineers to stop the measurements and adjust the suspensions to lower apparent viscosities. These are good procedures to follow. Process corrections should be made well before hard viscometer limits are reached and viscometers are damaged.

Dilatant and rheopectic suspensions can also severely tax and ruin mixers and pumps. It will almost certainly be obvious, if a ceramist is paying attention, when dilatant suspension rheologies are causing problems. Adjustments should be made quickly when such properties appear.

Professor Funk frequently told the story[8] about the student who ruined a set of ceramic ram press dies because the student wondered what would happen if he tried to press an extremely dilatant forming body. The student reasoned that if he first asked Professor Funk what would happen, he'd be told to "go try it." So instead of asking first, he tried it first. He only reported his experiment after the die was broken into pieces.

Professor Funk also often jokingly suggested to students that if they were mixing extremely dilatant suspensions on a milkshake mixer, or if they were measuring viscosities on a viscometer, and they noticed the motors beginning to smoke ... that was a good indication they were dealing with dilatancy. He joked about it, but it is true.

Rheopexy and dilatancy can both reach extreme levels. It's best not to subject a good viscometer to such suspensions if it's known in advance that they will exhibit extreme properties. Precise measurements on dilatant and rheopectic suspensions are difficult to achieve unless the effects exhibited are relatively mild.

For all these reasons, measurements demonstrating rheopectic properties are difficult to acquire and are, therefore, rare. When rheopexy happens, however, it will be the result of particle/particle interactions and collisions within the suspension.

Gelation Is Not Rheopexy

When apparent viscosities are monitored at low shear rates as gel structures form, the measured rheograms will look like the rheopectic curves in Figure 5.1. **Gelation,** however, **is not rheopexy**.

Such rheograms occur when measurement shear rates change (decrease) abruptly, for example, from $100s^{-1}$ to only $1s^{-1}$ and the viscometer continues to measure apparent viscosity versus time. After such a change in shear rate, gelation causes structures to rebuild and measured apparent viscosities to increase until the new gel breakdown/buildup rate equilibrium has been achieved. This is an example of an increase in measured apparent viscosity that accompanies a **decrease** in shear intensity.

In rheopectic suspensions, **increases** in shear intensity cause increased rheopectic behavior and increased measured apparent viscosities. The effects of such behavior are opposite to the effects of high shear on gel structures. High shear intensities break gel structures and reduce apparent viscosities in thixotropic suspensions. High shear intensities build structures and increase apparent viscosities in rheopectic suspensions.

Summary

The two time-dependent rheologies that ceramists must deal with are *thixotropy* and *rheopexy*. Thixotropy occurs as gel structures are broken down over a period of time by constant shear. As measured apparent viscosities of thixotropic fluids decrease with time, they approach limiting, minimum values. Rheopexy occurs as collision intensities increase over a period of time at constant shear. Measured apparent viscosities of rheopectic fluids increase with time as they approach limiting, maximum values.

In thixotropic fluids under extremely high shear conditions, all particles flow as individuals. In rheopectic fluids under extremely high shear conditions, all particles can be mechanically bound together into a single compact which can cause flow to cease.

Time-dependent rheologies respond to shear*time history. As the shear history increases, apparent viscosities of time-dependent fluids and suspensions can increase or decrease. When equilibrium has been achieved, apparent viscosities will remain constant as shear history continues to increase.

When the apparent viscosities of fluids are totally unaffected by shear history, those fluids are **not** time-dependent fluids. Newtonian fluids, for example, are not time-dependent.

Chapter Six

Attractive Forces and Gelation

When interparticle attractive forces dominate within particle/fluid suspensions, gelation takes place. The inter-relationships between these two phenomena will be covered in this chapter.

Attractive Interparticle Forces

Electrostatic Surface Charges

When particles are suspended in fluids, especially polar fluids like water, their surfaces can be expected to exhibit electrostatic charges. Deflocculating chemical additives can be used to enhance net charge densities to higher positive or higher negative values. Flocculating chemical additives can be used to cancel electrostatic surface charges.

The mineralogical types of particles, suspension pH values, and the concentration and type of additives all affect the sign and charge density of electrostatic surface charges.

Each different mineralogical type will have an *isoelectric point* (IEP). The IEP is the pH at which the net surface charge on a suspended powder is zero. At other pH values, additives containing ions of the opposite charge to the prevailing surface charge can be used to cancel and lower the net surface charge towards zero.

In a suspension containing particles with net negative surface charges, positively charged ions will be attracted to the negative sites (since opposite charges attract.) At the surfaces, the cations will be loosely bonded to the particles due to the electrostatic charge attractions. Relative to the larger volume of suspension, they will cancel the negative sites and lower the net surface charge towards zero.

Van der Waals Forces

There are two types of attractive forces within suspensions: electrostatic attractive forces between oppositely charged particles and ions, and van der Waals forces.

The attractive nature of van der Waals forces results from the attraction of the negatively charged electron clouds surrounding each atomic nucleus to the positively charged nuclei of other atoms.

Van der Waals forces of attraction are always present between atoms. They can't be turned off, removed, or eliminated. But they are only weak forces. Because they are weak, they're easily overpowered and masked by the presence of other attractive and repulsive forces. They are, nevertheless, always present.

When all other attractive and repulsive forces have been eliminated, van der Waals forces take over. Since they are weak attractive forces, they pull particles together slowly. The weak attractive nature of Van der Waals forces is responsible for producing flocculating conditions in particle/fluid suspensions.

Highly charged flocculating cations, such as Mg^{++}, Ca^{++}, Al^{+++}, etc., can also cause flocculation when particle surfaces are electrostatically negative. Such cations appear to serve two functions: their highly positive fields draw the negatively charged particles together, and their positive charges locally neutralize negative surface charges, thereby allowing van der Waals forces of attraction to function.

In the absence of repulsive interparticle forces in suspension, attractive forces pull particles together to create flocs which grow into large, continuous gel structures that reach throughout the whole volume of suspension. This process is called both *flocculation* and *gelation*. Under the influence of attractive forces, individual particles come together to form small *flocs*. Small flocs grow into larger flocs, and larger flocs combine to form large 3-D *gel structures*.

Flocs and Gel Structures

A floc is a small group of particles, loosely associated by attractive forces. Figure 6.1 shows a floc structure. Particles in such structures are weakly bound together by electrostatic and/or Van der Waals forces. We're all familiar with dry floc structures that take the form of fuzz balls (a.k.a. dust bunnies, belly button lint, etc.)   Floc structures within particle/fluid suspensions are similar.

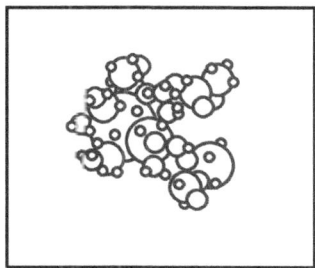

Figure 6.1 A Floc

Flocs are not strong, chemically bonded structures. Particles are loosely attached to one another As a result, flocs can easily break apart when sheared.

Some agglomerates, such as calcined alumina granules, have gone through a firing operation. Sintering and/or melting phenomena can cause closely associated particles to react with each other and grow together to form agglomerates. Such agglomerates will be strong because the particles are chemically bonded to one another. To break apart strong agglomerates and free the individual constituent particles, one must use intense mechanical impact forces. Once such agglomerates are broken, constituent particles will not reform as strong agglomerates without going through another firing operation like the one that produced the original high-strength agglomerate structures in the first place.

Flocculated particles and gel structures, however, are **not** chemically reacted, fused, sintered, nor strongly bonded as are the agglomerates in the preceding example. Flocculation always produces structures that are held together with relatively weak electrostatic and/or Van der Waals forces. Flocculated structures in suspensions will break apart when they are subjected to relatively low shear intensities, such as during stirring, mixing, or pumping. When shearing actions stop, floc structures will again form.

Figure 6.2 shows examples of two particles that are flocculated by interparticle attractive forces, and two other particles that are strongly bonded by a sintering operation. The purpose of this example is to show that floc structures (and their larger gel structure counterparts) are **not** the same as strong, sintered agglomerates.

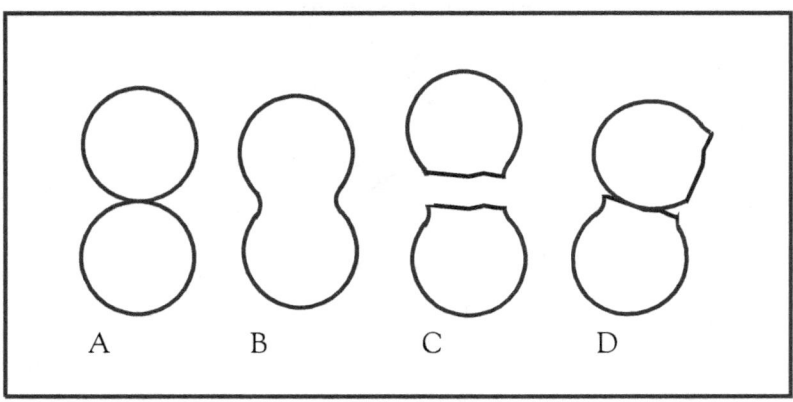

Figure 6.2 Bonding and Rebonding in Flocs and Agglomerates:
a Floc (A), a Sintered Agglomerate (B), the Agglomerate after
Milling (C), and the Flocculated Milled Agglomerate (D).

Figure 6.2A shows two flocculated particles held together by weak Van der Waals forces. Under shear, the two particles can separate and travel as individuals, but when the system returns to a quiescent state, the particles can again flocculate as in (A).

Figure 6.2B shows two particles that have been sintered to form an agglomerate which functions as a single particle. Shear during mixing or pumping may not be sufficient to break such strong bonds.

By comparison, the floc shown in Figure 6.2A consists of two particles. The two particles are weakly bonded to form a single floc, but the floc consists of two particles, nevertheless.

When the agglomerate in Figure 6.2B is broken by milling, for example, the two constituent particles are liberated, as shown in Figure 6.2C. The particles in Figure 6.2C are free to travel independently of one another during shear.

When shear stops, and if the particles are in a flocculated suspension, attractive forces can pull them together in the form of a floc as shown in Figure

6.2D. They will not revert to their original agglomerated form shown in 6.2B. The agglomerate will not reform without another sintering operation. In the form of 6.2D, the two particles can separate under shear, and reflocculate when the system is quiescent.

As the flocculation process takes place, individual particles come together to form small flocs, and the small flocs and other individual particles continue to grow into large 3-D gel structures. It may be helpful to consider gel structures to be in the form of large, 3-D fishing nets.

3-D Fishing Nets

Consider a fishing net, which has been defined as "a bunch of holes tied together with string."[8] To be useful, fishing nets are two-dimensional.

But consider the extension of the fishing net's 2-D structure into three dimensions, to achieve a three-dimensional fishing net. Such a 3-D structure would actually be useless as a fishing net, because it would be impossible for fish to enter. But it is an excellent picture of the structure of a 3-D gel.

The knots must be positioned randomly, but uniformly, throughout the whole volume occupied by the 3-D net and all knots must be connected with strings to their nearest neighbors. Each knot and each string in the 3-D net represents groups of flocculated particles bonded together in gelled structures similar to the one shown in Figure 6.1.

The full range of gel structure possibilities can also be demonstrated using 3-D fishing nets. Some nets could be formed using small strings (with small knots). Others could be formed with large diameter ropes (with correspondingly large knots). Some nets could have relatively long strings or ropes between knots. Others could have relatively short strings or ropes between knots. The full range of gel structures can be simulated by these types of variations.

The natures of the holes and channels through such nets are of particular importance to the subject of gel structures. Note that the sizes of the holes, the volumes of the channels, and the ease with which fluids can move through such systems are defined by the particular characteristics of the string (or rope), and the volume density of knots that form the fishing net. The ease of fluid motion within gel structures similarly relates to the characteristics of the channels throughout the flocculated structures that form the gel.

3-D fishing nets are excellent, but simplified, pictures of actual gel structures that form in flocculated suspensions.

Flocculation

Within suspensions, flocculation produces 3-D gel structures of particles that are very similar in form to the 3-D fishing nets just described. Instead of knots, gel structures contain large flocs of particles. Instead of strings, gel structures contain flocculated chains of particles that connect the flocs. Within an actual gel structure, it would be difficult to distinguish between the flocs (corresponding to the knots in a net) and the chains (corresponding to the connecting strings), but the overall picture is accurate.

The key to understanding gel structures is to realize that they are large, 3-D structures that extend throughout the whole volume of suspension.

Many believe that flocculation produces only small flocs such as the one shown in Figure 6.1. This is partially accurate because flocculation does start with small flocs. But flocculation in high solids content ceramic suspensions produces large structures that grow well beyond the initially formed, small flocs. Allowed to proceed to completion, flocculation produces large 3-D gel structures.

Colloidal impurities in water purification plants won't settle easily as individual particles. Flocculating chemicals are added to help draw them together to form flocs (such as in Figure 6.1) that are large enough to settle and be removed during filtration. The solids contents in water purification systems, however, are very low by comparison to the solids contents in typical ceramic suspensions. At the low solids contents in water purification systems, a small floc, such as the one in Figure 6.1, is the appropriate picture for flocculation behavior. Figure 6.1, however, is not an appropriate picture for the **overall** gelation process in ceramic suspensions.

The Gelation Process

When shear intensity is sufficient, all particles will be freed from an existing gel structure to report as individuals. When the stirring intensity is reduced, attractive forces will begin to pull the particles together again and small flocs will form. As gelation continues, more particles will join the growing structures. Some flocs will increase in size; other small flocs will form; and some flocs will connect to other flocs as new particles fill the gaps between them. Individual particles will continue to join the growing structure until all free particles have been incorporated. Any remaining independent flocs will continue to join the larger structure until all free flocs have been incorporated into one overall structure.

When allowed to grow to completion, all particles will be part of one large gel structure that fills the whole volume of suspension. In a completed gel structure, no individual particles will remain independent of the structure.

Gel structures are characterized by (1) large continuous floc structures connected by flocculated chains of particles and (2) continuous fluid channels within and throughout the structure. Suspension properties and applied shear conditions determine when the structures have reached their equilibrium arrangements. After all

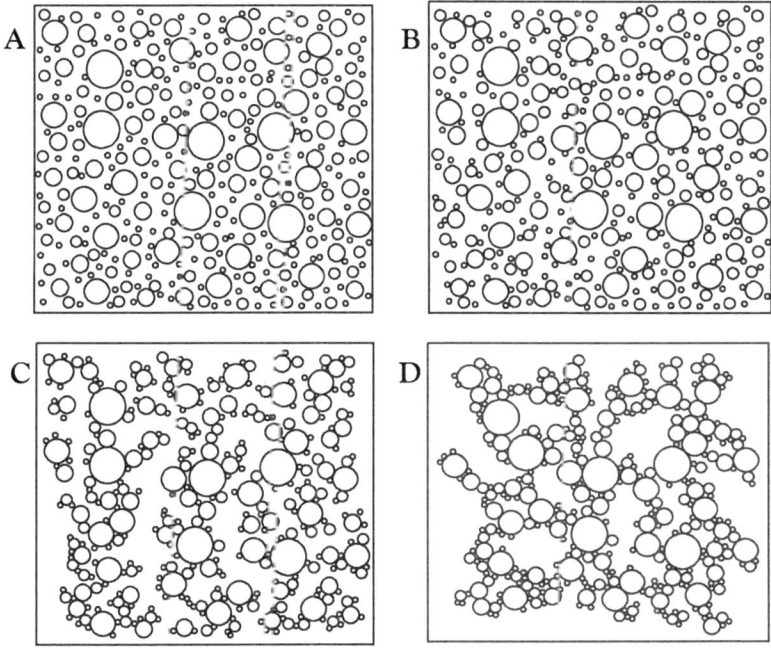

Figure 6.3 Example of the Gelation Process

particles and all flocs have been incorporated into a single, large gel structure, strong interparticle attractive forces can continue to densify and strengthen the structure and expel fluid into the channels. This is known as *syneresis*.

Figure 6.3 shows a series of diagrams that demonstrate the gelation process with time from A to D. A reminder: these diagrams show 2-D depictions of the 3-D gelation process.

Figure 6.3A depicts a suspension of particles, all of which are independent of one another. Figure 6.3B shows some small flocs beginning to form within the overall array of individual particles. Figure 6.3C shows the initial stages of growth of the floc structures. Some small flocs are still independent of the large structures, but most individual particles have been incorporated into the growing structures. Fluid channels begin to appear between the flocs. Figure 6.3D shows a completed structure. All particles and flocs in 6.3C are part of the large, single structure in 6.3D.

Note that all particles, and especially the finest particles, are incorporated into the completed gel structure in Figure 6.3D. No individual particles are free to travel independently. This is an important property of flocculated suspensions and gel structures. All particles (especially the colloids) are tied into the structure and immobilized.

Summary

Attractive interparticle forces produce gel structures in suspensions. Van der Waals forces, with help from electrostatic attractive forces, are responsible for gelation and flocculation behaviors. When time and shear conditions allow, gelation produces large 3-D gel structures that extend throughout the whole volume of a suspension.

Within ceramic suspensions, the picture of a *floc* as a small group of particles is accurate only as one of the initial stages of the flocculation process. The correct picture for flocculation in high solids content ceramic suspensions is the production of a complex, 3-D gel structure that extends throughout the suspension.

Chapter Seven

Shear-Thinning Rheologies

When gel structures are present in suspensions, shear-thinning rheologies can be expected.

Shear-Thinning Behavior

Gel structures typically produce shear-thinning rheologies. This occurs because the imposed shear causes gel structures to break down and this lowers apparent viscosities.

When a flocculated suspension is quiescent, the gel structure will build to form a complete structure that extends throughout the whole suspension, such as the structure shown in Figure 6.3D. Flocculated particle/fluid suspensions containing completed gel structures can be expected to exhibit yield stresses. They are examples of *yield shear-thinning*, or *yield-pseudoplastic rheologies*.

When a gel structure is sheared, for example by mixing or pumping, the large gel structure will break down. The intensity of shear will determine the size of the flocs that flow free of the gel structure. Low intensity shear will produce relatively large flocs. At the other extreme, high intensity shear can destroy flocs and cause all particles to report as individuals so each particle can flow independently of all others.

Apparent viscosities will be relatively high when measured under low shear conditions in which large flocs are flowing. Apparent viscosities will be relatively low when measured under high shear conditions in which small flocs or individual particles are flowing. The gel breakdown phenomenon explains the behavior of shear-thinning suspensions.

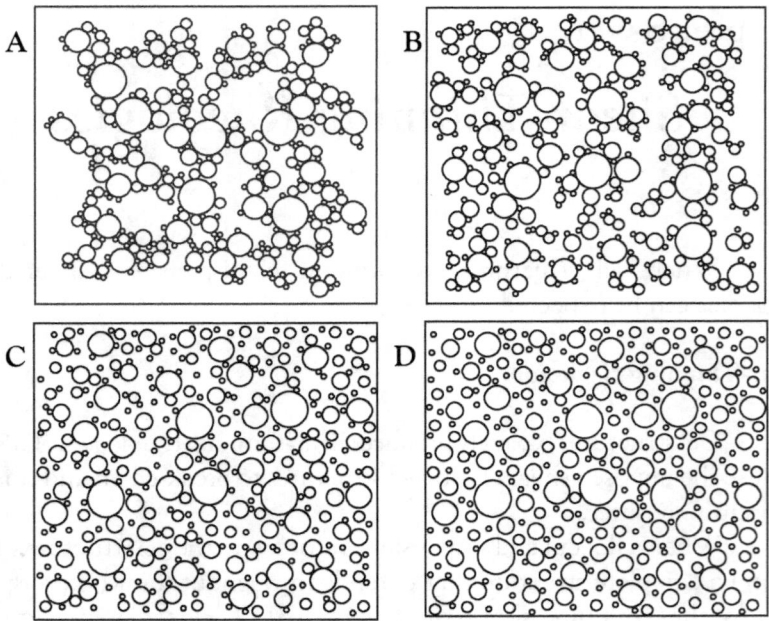

Figure 7.1 Example of the Shear-Thinning Process

As mentioned earlier, gelled suspensions exhibit dynamic equilibria. As shear destroys gel structures, attractive interparticle forces attempt to rebuild those structures. When the rates of gelation and breakdown balance, the systems will exhibit constant apparent viscosities during shear. This is the reason why suspensions flowing in pipes achieve steady-state viscosities and flow conditions.

Figures 7.1A through 7.1D demonstrate what happens when a gel structure is sheared. Figure 7.1A is a picture of the quiescent gel structure. Figure 7.1B is a picture of the structure under low shear conditions; Figure 7.1C is a picture of the structure after shear at higher rates; and Figure 7.1D corresponds to high intensity dispersion conditions when each particle reports individually within the suspension.

High intensity dispersion (HID) conditions, defined as mixing with impeller tip speeds greater than or equal to 5000 ft/min, should completely destroy all gel structures and set all particles free to travel independently. In fact, high intensity dispersion conditions should strip most, if not all, adsorbed ions and chemical additives from the surfaces of the powders and force them out into the *interparticle soup*. Under HID conditions, the *interparticle soup* will contain individual particles, ions, and molecules.

When HID conditions are removed, everything in the suspension will be free to move towards and take equilibrium positions. Ions and additive molecules will adsorb onto particle surfaces in equilibrium positions, the particles will reflocculate, and the gel structure will rebuild.

Lower intensity shear rates should not be expected to destroy all of the gel structure. The nature of the attractive forces within a suspension and the strength of its gel structure will determine how intense the shear must be to cause all particles to report as individuals.

As soon as gelation rates exceed the breakdown rates produced by the imposed shear, gel structures will again grow and/or strengthen.

Thixotropy

The breakdown of gel structures also explains time-dependent thixotropic fluids and suspensions. Just as increased shear rates cause gel structures to break down, constant shear rates applied for even short periods of time will also cause gel structures to break down.

Shear-thinning time-independent rheologies are defined by instantaneous changes in apparent viscosities as shear rates increase. In practice, however, gel breakdown doesn't occur instantaneously. Most shear-thinning suspensions are therefore also thixotropic. The thixotropy appears as gel structures continue to break down with time until they achieve buildup/breakdown equilibrium at the imposed shear rates.

When process suspensions are flocculated, one should expect to see both shear-thinning and thixotropic behaviors. Because both of these are caused by gel breakdown, they occur together.

Summary

Yield shear-thinning (a.k.a. yield-pseudoplastic) and thixotropic rheologies are characteristic of flocculated particle/fluid suspensions. The apparent viscosities of these suspensions will decrease as shear is applied, and increase again as gel structures rebuild after shear is removed.

The rheologies of flocculated and partially flocculated ceramic suspensions are **yield** shear-thinning (rather than simply shear-thinning) because the gel structure which produces the yield stress is necessary to allow formed wares to hold their shapes. Pure shear-thinning fluids (without yield stresses) are not only rare, but unimportant within ceramic process systems.

The rheologies of flocculated and partially flocculated ceramic suspensions are also thixotropic for the same reason. It is doubtful that ceramic suspensions that do not exhibit yield stresses would ever show indications of thixotropy.

Gel structures, yield stresses, yield-shear thinning behaviors, and thixotropy are all closely related. If measurements show the presence of any one, the others should also be present.

Chapter Eight

Particle/Fluid and Particle/Particle Interactions

Particles traveling within fluids interact with the fluids and with other particles according to their physical properties and the nature of the carrier fluid. Discussions in this chapter will focus on these types of interactions.

Particle/Fluid Interactions

Two Examples

Particles interact with their carrier fluids. When the masses of particles are sufficiently small, they will be carried along with their carrier fluids. Massive particles behave relative to their carrier fluids just as the proverbial 500 lb gorilla behaves. They do their own thing.

Consider a bowling ball. If one is dropped to the ground from a 3' height, most of us can predict where that ball will hit the ground. It's doubtful that any breeze (short of hurricane force winds) would have much effect on the landing site of the ball.

If the bowling ball is dropped from a 3' height into a swimming pool, the velocity of the ball as it hits the bottom of the pool will be considerably less than the velocity with which it hit the water. This difference is directly related to the properties of air versus the properties of water. A bowling ball will behave differently when it falls through water than it will when falling through air.

Now consider tiny particles of clay. If they are dropped from the same 3' height, their landing sites are unpredictable. Such particles are so small that even a slight breeze can cause them to move sideways and not drop straight down as did the bowling ball. Strong breezes would carry such particles away into the atmosphere. Where they land then would be anybody's guess.

If a few clay particles were dropped into the water in a pool, they would probably float for a while on the surface due to surface tension and to their lack of mass. Were they heavier, they would be expected to quickly pass through the surface and continue to the bottom of the pool. Even though the density of clay is greater than that of water, clay particles may not sink until agitation causes their surfaces to be completely wetted with the water.

Even then, when the clay particles are under the water, they still may not sink because the flowing water can drag them along with it. Water is more dense than air, so flowing water will have more influence on the particles than air currents do. If the particles are small enough and the water is quiescent, Brownian motion may still prevent them from settling.

These two examples show some of the differences that can occur when particles of different sizes interact with fluids.

Several Considerations

For a particle to be affected by the carrier fluid, the mass of the particle must be small enough so the momentum of the flowing carrier fluid can accelerate the particle in the direction of flow. When a horizontal breeze encounters a falling bowling ball, as in the example above, the momentum of flowing air molecules will be minuscule and insignificant relative to the momentum of the falling bowling ball.

The momentum of flowing air molecules, however, is substantial relative to a tiny clay particle. Even in a gentle breeze, the clay particle will be accelerated in the direction of the air flow, and depending on the size of the clay particle and the stiffness of the breeze, it may be carried off completely.

Flowing water will have even more influence on the clay particle because water is denser and flowing water will have more momentum than flowing air at the same velocity. Flowing water will have more influence on the falling bowling ball as well, but due to its size and mass, the bowling ball will still generally head in the direction its momentum and the acceleration of gravity carry it. Due to the large mass difference, the bowling ball will experience only minor changes to its momentum from interactions with flowing water.

We could now head to the physics and chemical engineering textbooks to show the applicable equations that govern these phenomena, but the specific equations are not important to this discussion. I'm sure many of you are now relieved, saying, "Whew!!" and wiping the sweat from your foreheads.

The author has never actually had to calculate the velocities and interactions between particles and fluids in suspensions. Ph.D. students may need to perform such calculations for their theses, but ceramists in industry do not need to perform such calculations. It is sufficient for ceramists to recognize, consider, and understand the possibilities of such interactions, and the influences they have on the behaviors of suspended particles.

Consider also two bits of information garnered from sedimentation particle size analysis technology: (1) Colloids, which are defined as particles smaller than 1μm, do not settle as the Stoke's Equation predicts because they are affected by Brownian motion; and (2) Particles larger than ~70 - 100μm usually settle quickly through suspensions. When they do, they can affect and alter the settling rates of other smaller particles in their vicinity.

This information indicates that settling phenomena do not usually apply to colloidal particles. On the other hand, settling can occur quickly for most particles in the sieve size ranges.

Most ceramic suspensions contain colloidal particles. Traditional ceramics suspensions usually contain particles that range in size from colloids to greater than 100μm. Within electronic ceramics suspensions, anything larger than ~20μm is considered to be a 'boulder.' By this definition, refractory suspensions frequently contain large percentages of 'boulders.' So within the various branches of ceramics, particle sizes can run the range from colloids to several centimeters in diameter.

Carrier fluids can be water, a variety of organic fluids, or air (pneumatic conveying systems.) All of these are light molecules that will have relatively minor effects on suspended particles. Colloidal particles and slightly larger fines will be most influenced by the flow of carrier fluids. Particles that are 70μm or larger will be least influenced

Colloids

Generally speaking, colloidal particles will travel with the carrier fluids unless they are immobilized in some way. Their masses are small enough that the carrier fluids can exert major influences over their directions of travel.

Consider water that flows around a cylindrical thermometer that has been inserted perpendicular to the water's direction of flow in a pipe. As the water flows down the pipe and encounters the thermometer, it will flow around the obstruction (the thermometer) and continue down the pipe. When colloidal particles are suspended in the water, they will be accelerated around the thermometer by the water, and they will continue down the pipe relatively unhindered by the obstacle. Such particles will typically flow with velocities that are very similar to, and usually equal to, that of the flowing fluid. When the fluid takes the detour to flow around the thermometer, so will the entrained colloids.

Only at really high flow velocities will the differences in mass between the colloids and that of the carrier fluid begin to show differences between the momentum of the fluid and the momenta of the colloids. Generally speaking, fluids don't flow fast enough in pipes for this to happen to colloids.

Coarse Particles

At the other extreme, if one considers $100\mu m$ particles to be an 'extreme,' flowing fluids may not have sufficient momentum to fully entrain the particles. With time and straight flow, $100\mu m$ particles will usually also travel with velocities similar to the carrier fluids. But when an obstruction appears, the fluid may not have enough momentum nor sufficient influence over the larger particles to accelerate them out of the way of the obstruction. When this occurs, the fluid will flow around the obstruction, but many of the particles flowing in line with it will collide with it. Some particles may form a deposit on the leading edge of the obstruction and some particles will bounce off the obstruction (possibly abrading it in the process), flow around it, and continue down the pipe.

Remember: Momentum is mass times velocity. The capability of the fluid to influence particle velocities depends upon whether the fluid has enough momentum, due to its combined mass and velocity, to affect the particles' momenta. When the particles are large enough, carrier fluid momentum will have little effect and the particles will collide with the obstruction, even though the carrier fluid flows around it.

The Effects

What common examples of particle/fluid interactions have we all seen? How does such behavior present itself in ceramic process systems? What problems face ceramists when such phenomena occur?

Bugs on Windshields

This seems like a weird subject, but it is a common example of how particles (bugs) interact with fluids (air) and with passing automobile windshields (obstructions).

Have you ever noticed that most fancy sports cars have few bugs marking their windshields? The aerodynamics of such cars are finely tuned so they can run at high speeds with minimal wind resistance. Air flows very smoothly over and around such designs.

Tilt-cab trucks with large, flat frontal surfaces are at the opposite extreme from the fine sports cars. Wind resistances are substantial for such trucks.

While it is relatively easy for air to flow around a sports car, it is also relatively easy for bugs in the air stream to flow with the air stream around the sports car. Little momentum is needed to move the bugs out of the path of such cars. The only bugs that typically will hit the windshield of a sports car will be the really large, slow-moving varieties. This also applies to many of today's family cars, which are very aerodynamic compared with most of the family cars from the 20th century. Small bugs and mosquitoes simply don't hit the windshields of most of today's cars. But big bugs do.

The windshields of tilt-cab trucks, however, will be hit by much larger assortments of bugs, and especially by smaller ones. It will be difficult for bugs aligned with the center of such windshields to be accelerated out of the way by the air stream as the trucks pass. Since such trucks present large frontal areas, bugs ahead of the trucks will have to be accelerated several feet laterally to avoid collisions. Because such trucks are not very aerodynamic, even smaller bugs will collide with them.

Everyone should have some experience with the phenomena in this example. The next examples, however, are more engineering-oriented.

Isokinetic Sampling

Interactions between particles and carrier fluids are at the center of isokinetic sampling procedures used in stack sampling systems. When stacks are sampled to measure particle size distributions and the amounts of particulates discarded into the atmosphere by industrial exhaust gas streams, the sampling must be done isokinetically. This means that the velocity of gases entering the tip of the sampling probe must exactly match the velocity of gases at the sampling point in the exhaust gas stream. When they match, gases and entrained particles flow smoothly into the test probe. This allows accurate sampling and accurate calculation of the mass of particulates exhausted into the atmosphere. When sampling is not isokinetic, and the velocities in the flue and entrance to the sampling probe differ, test results will not accurately characterize the exhaust stream.

Figure 8.1 shows three different examples of sampling. Figure 8.1A shows isokinetic sampling where the velocity within the probe is equal to the velocity in the exhaust gas stream. Figure 8.1B shows sampling when the velocity within the probe is less than the exhaust gas stream velocity, and Figure 8.1C shows sampling when the velocity within the probe is greater than the exhaust gas stream velocity.

When an isokinetic sampling probe is inserted into a stack, the goal is to have all gases and particles traveling within the column of fluid aligned with and defined by the sampling probe nozzle's outer circumference to smoothly flow into the nozzle. No more. No less. This is shown in Figure 8.1A.

Figure 8.1B shows conditions that would occur if insufficient suction was applied to the probe to match the exhaust gas stream velocity. In this case, even the particles directly in line with the probe could flow around it with the gas stream. When the gas velocity within the nozzle entrance is less than the exhaust stream velocity, the nozzle will act like a blunt object obstructing exhaust gas flow. Gases and small entrained particles will flow around it. Only large particles flowing directly towards the probe could impact the slower velocity gases in the nozzle entrance, enter the probe, and be sampled. This would produce an incorrect sampling result: the quantity of particles (and especially the fines) flowing in the exhaust gas would be grossly underestimated.

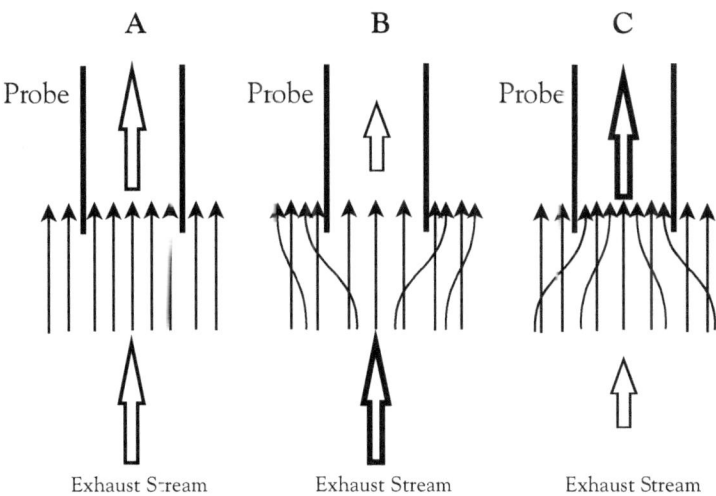

Figure 8.1 Stack sampling examples: (A) Isokinetic,
(B) Probe velocity too low, and (C) Probe velocity too high;
Hollow arrows signify probe and exhaust stream velocities;
Solid arrows are coarse particles; and dashed arrows are colloids.

If full vacuum from a large pump was applied to the probe, the probe could act like a vacuum cleaner and suck most of the particles from the stack, regardless of the particles' positions in the flow stream relative to the probe.

Figure 8.1C shows conditions that would occur when the suction on the probe was too great and the gas velocity within the probe was high, relative to the exhaust gas stream velocity. Lots of fine particles would be pulled from the stack into the probe (especially the colloids), but the coarsest particles in the stack would hardly be influenced. Results would again be in error because the calculation is based on the assumption that all particles entering the probe were traveling in line with it in the stack. In this case, the mass of particles in the exhaust gas stream would be severely overestimated.

Only when the exhaust gas velocity and the velocity in the probe entrance are identical (i.e., isokinetic, as in Figure 8.1A) will the sampling and results be accurate. When the two velocities are identical, any gases and entrained particles that are flowing directly towards the nozzle's entrance will flow smoothly into the nozzle, be captured in the probe's filter, and be successfully sampled.

Milling and Mixing

In milling and HID mixing systems, where the goal is for the milling media and mixing blade tips to impact particles and agglomerates, particle/fluid interactions are important.

When particles are so small that they easily flow with the carrier fluids, they can miss the milling and mixing events they are intended to see. When solids contents are too low and particle sizes are too small, milling and high intensity dispersion can both be inefficient or ineffective.

For example, it is extremely difficult (nearly impossible????) to mill all particles in a standard, large production ball mill to achieve 100% finer than $\sim 10\mu m$. If it was necessary to do so, milling times would be very long and even then, desired finenesses might not be achieved.

The largest particles in such mills have the greatest probability of being impacted and broken by a milling event. Probabilities of impact decrease as particle sizes decrease. The reason why the largest particles see the most impacts is that they are the biggest targets and they are the least influenced by the carrier fluids.

As milled particles decrease in size, they are more easily entrained to flow with the carrier fluid and they can be accelerated out of the way of impending milling events. In air swept ball mills, the velocity of the air traveling through such mills is controlled so it will entrain all particles below a certain size and sweep them out of the mill. In this way, no energy is wasted by milling particles any finer than necessary. As soon as the particles are small enough, they are swept out of the mill.

In dry batch mills when particles are small enough, they can stay entrained in the air streams and miss all further milling events. In wet ball mills that are run at low solids contents, the fines can be accelerated out of the way of impending milling events for similar reasons.

Another reason why fines are not easily impacted deals with their sizes relative to media sizes. When milling media can be as large as 2", 3", and 4" diameter balls, and fines are $20\mu m$ or less, chances of media exactly impacting fine particles are slim. How many fine particles will be exactly in the impact area as two milling media balls collide? If the impact hits a larger particle first, smaller particles within the impact zone may be shielded from the full force of the impact.

The inherent strengths of particles usually increases as particle size decreases. Coarse particles will preferentially fracture along flaw lines. As particles are reduced in size, fewer and fewer flaws remain, and the inherent particle strengths increase. Flawed (and inherently weaker) coarse particles are more likely to see the full energy of media impacts, while the stronger, less-flawed fines are less likely to see the full energy of media impacts.

There is no easy solution to this problem in a dry batch mill or in an air-swept continuous ball mill. One solution is to use different types of mills that have been designed for efficient comminution of particles in these finer size ranges. Fluid energy mills, for example, could be used.

In wet milling systems, one excellent solution is to crowd the particles. If there is no adjacent space into which a particle can move to miss an impending milling event, the particle will see the impact and be reduced in size. One way to crowd suspended particles is to raise the solids content of the feed suspension. Keep in mind, however, that raising the solids content in a ball mill will change the particle size distribution of the product powder.

The other solution which is valid for wet systems is to use a different type of mill. Vibratory ball mills and stirred ball mills are commonly used to wet mill fine powders.

High intensity dispersion (HID) systems rely on impacts between the teeth on the high speed dispersion blades and the flocs, particles, and agglomerates to perform deagglomeration and delamination. When low solids content suspensions are used, HID doesn't properly deagglomerate or delaminate. At low solids contents, particles can be moved out of the impact zones by the carrier fluids so they will not be properly impacted by the blades. When this happens, energy is wasted.

Suspensions must be reasonably crowded for HID to work properly. Systems that are too crowded can be inefficient as well, but some crowding is necessary.

Temperature is a good indicator to tell whether HID is running efficiently. When room temperature suspensions are fed into HID systems and they quickly reach $\sim 70^\circ C$, the HID is being successfully accomplished. In Continuous HID (CHID) systems, this can happen with dwell times in the CHID chamber as short as 30 seconds.

Another reason why crowding enhances milling and HID phenomena is that in crowded systems, impact forces will be transmitted from particle to particle within the impact zones. The probability that two milling balls will exactly impact a single fine particle is relatively low. But the same two milling balls will have a high probability of impacting several of the particles that lie between them in a crowded suspension.

To use a football analogy, a single running back may have an easy time eluding a single defender between himself and the goal line. But that same running back will have a much more difficult time running through the center of the zone in which all of his teammates and the whole opposing team are crowded. He may bounce off of other players frequently as he tries to pick his way through the pack and move up the field. Even when none of the defenders are free to tackle him, the holes can close, and he can be stopped due to the crowded field around him.

Consider also what happens when someone fumbles the ball. One or two players dive on the ball and everyone else dives on top of them, forming a huge pile of players on the field. The player(s) on the bottom of the pile will feel the impacts of everyone who dives onto the growing pile. Impacts will be transmitted from player to player from the top down and through the pile to the guy(s) at the bottom with the ball.

Similarly, this type of phenomenon happens in milling and HID systems. Individual particles may not physically come in contact with the milling media or the mixing blade, but when the system is crowded, impacts will be transmitted from media to particle to particle ... to particle to media. As a result, many particles will see the impact events, even though only relatively few may actually be in contact with the media or mixing blades.

The person(s) in charge of milling and HID operations should consider whether the conditions being used favor the impacting of the desired particle fractions in their suspensions. If conditions do not favor the desired impacts, the operators might want to change suspension properties, or milling or HID operating parameters, or they might want to use a completely different type of mill, to successfully achieve their product materials.

During the process of making such decisions, considerations should especially be given to whether the particles are sufficiently free to miss the comminution and HID events, or whether the suspensions are sufficiently crowded so the particles cannot escape these events.

Filter Pressing and Casting

Immobilize the Colloids

During dewatering operations such as filter pressing and slip casting, particle/fluid interactions are again important. When the suspended particles are not quickly immobilized by flocculation and gelation phenomena, they will be influenced by the flow of the carrier fluids. This is particularly applicable to colloids which can easily be influenced by flowing fluids.

When suspensions are fully or partially flocculated, colloids will be quickly immobilized by incorporation into the growing gel structures. When suspensions are fully or partially deflocculated, colloids can remain free and mobile where they are easily influenced by the flowing carrier fluids.

In many (all???) ceramic processing plants that perform slip casting and filter pressing operations, the goal is usually to cast or press as quickly as possible. When these operations are controlled to proceed quickly, the water (the carrier fluid) will flow with relatively high velocities in the direction of the filter surface. When deflocculated suspensions are used, the flowing fluids will drag any free particles towards the filter. The fluids will exert greatest influence on the finest particles (the colloids) and that influence will diminish as particle sizes increase.

As a result, a filter press suspension can be unmixed with the particles segregating by size from the smallest at the filter cloths to the coarsest at the centers of the filter cakes. The particle size distribution of the whole suspension may pack reasonably well, but when unmixing and size segregation occur, packed porosities will be high and cake permeabilities at the filter surface will be low. Because the colloids are most easily dragged along with the carrier fluids, the cake surfaces at the filters can be very smooth because of the colloids, and the surface at the center of the cake can be very coarse and rough. Many such cakes can be pulled apart into two halves as the coarse particles at the centers of the cake are easily separated.

The reason is simple to explain. When the finest colloids in a deflocculated suspension are dragged along with the flow until they reach the filter surface, the first layer of cake at the filter cloth will be predominantly colloids. This first layer will essentially be a monodispersion of the finest particles in the suspension. Monodispersions pack poorly. Interparticle porosities in monodispersions can be 40% and greater. But the effective diameters of the pore channels within this first, monodisperse layer of colloids will be small due to the small colloid diameters.

When all further fluid to be removed must pass through such a layer of colloids with extremely small flow channels, dewatering rates will decrease quickly, and filter pressing times will increase substantially.

Deflocculated suspensions should not be used in filter pressing and casting operations. Flocculated suspensions should be used. In flocculated suspensions, the colloids (and coarser particles) can be easily immobilized and incorporated into the forming gel structure. As gel structures form, fluid to be removed can flow through relatively large flow channels formed within the 3-D gel structure.

The key expression to remember is: *Immobilize the colloids*. This occurs in flocculated suspensions. When the conditions are right for colloids to be immobilized, dewatering can proceed quickly.

Conditions for High Dewatering Rates

Some ceramists have applied the wrong reasoning to guide them when performing filter pressing and slip casting operations. They reason that since higher solids content suspensions contain less water, they should perform filter pressing and casting at the highest possible solids contents if they want the operations to proceed quickly.

The emphasis in this reasoning is on the *amount* of water, but it should rather be placed on the *ease of flow* of the water. To achieve reasonable suspension viscosities when the solids contents are maximized, suspensions must be well-deflocculated. Dewatering of well-deflocculated suspensions will proceed slowly.

When suspensions are tuned so the flow channels throughout the gel present the least resistance to water flow, filter pressing and slip casting operations can proceed quickly. Such behaviors occur in lower solids content, flocculated suspensions. The colloids and other fines are quickly immobilized, the flow channels throughout the gel structure will be reasonably large, and the water can filter through the structure relatively easily.

The issue is not *how much water* must be removed, but *how easily that water can flow* through the forming filter cake. As it turns out, the lower the solids content and the more water to be removed, the higher the state of flocculation required, and the faster the dewatering operation will run.

Conversely, the higher the solids content and the less water to be removed, the higher the state of deflocculation required, and the slower the dewatering operation will run.

If this is difficult to believe, test it in the lab. It works. Set up several suspensions at different solids contents but with similar apparent viscosities. The series of suspensions should cover the range from low specific gravity, flocculated suspensions to high specific gravity, deflocculated suspensions. Filter press them in the lab and monitor the filtration rates and the cake buildup rates. The lower solids content, more highly flocculated suspensions will win.

Abrasion

Keep in mind that when suspension conditions are right for the coarse particles to collide with obstructions in the flow paths, those obstructions will be subjected to high levels of abrasion. A thermometer in a pipe, at the very least, will be slowly polished and abraded by particles in suspension. As pipe velocities increase, abrasion intensities increase, and abrasion levels can ultimately become similar to those achieved in sand blasting operations.

To minimize abrasion problems, particle sizes in suspension should be small, and solids contents and suspension velocities should remain low. All three of these favor conditions where particles are easily entrained by the fluids to flow around obstructions.

Abrasion in suspension systems is one of those phenomena that can never be eliminated. Proper controls may minimize its deleterious effects, but when dealing with suspensions, ceramists should anticipate the need to deal with abrasion.

Particle/Particle Interactions

Interactions between particles occur when particle momenta are too large for the fluid to accelerate them out of the way of impending collisions with other particles. In the preceding section, large obstacles were considered. Each particle is a potential (albeit small) obstacle for all other particles.

When two particles approach, each will be accelerated to the side by the fluid to miss the potential collision. When momenta are too high, the particles will collide. After collisions occur, the velocities and momenta of both particles can be in directions totally different from the fluid flow direction. As the numbers and intensities of particle/particle collisions increase, particles can begin to bounce from collision to collision which could initiate rheopectic behaviors and the onset of dilatancy.

The particles most likely to collide are the largest particles with the greatest momenta. As shear rates, particle velocities, and momenta increase with increasing flow velocities, smaller and smaller particles can also collide.

As discussed earlier, colloidal particles (unless immobilized) generally travel with the fluids. When an obstacle, in this case another particle, presents itself, the carrier fluid will flow around the obstacle. Colloidal particles entrained in the fluid can be easily accelerated to the side to flow around obstacles. But as particle momenta increase, particles will increasingly behave independently from the carrier fluid and particle/particle collisions can begin to occur.

Since particle momentum is a function of its mass and velocity, particles with large masses and/or large velocities can have sufficient momenta to travel independently of the fluid and collide with other particles. Colloids traveling with sufficiently high velocities, or coarse particles traveling at relatively low velocities can easily collide with other particles, even when the carrier fluid flows smoothly around the other particles.

Figure 8.2 shows examples of two particles approaching (A) and interacting (B and C). If the approach momenta are low and the fluid can sufficiently influence the two particles, they can be accelerated out of each other's paths to flow around each other as in (B). If the momenta of the approaching particles are large, however, the particles could collide as in (C). Note that when the particles do not collide, they can continue to flow in approximately their same original direction with the carrier fluid. When they collide, however, they will move off in new directions after their collision.

The interaction shown in Figure 8.2 (B) can occur when the particles in (A) are colloids, even when they are moving reasonably fast, or when the particles in (A) are ultra-slow-moving coarse particles. Collisions as in (C) are more typical of coarse particle interactions, although they can also occur between high velocity colloids.

The natures of the collisions as in (C) will depend on the particle material characteristics. When particles have relatively smooth surfaces, or when they are coated with polymers, collisions may include some surface sliding

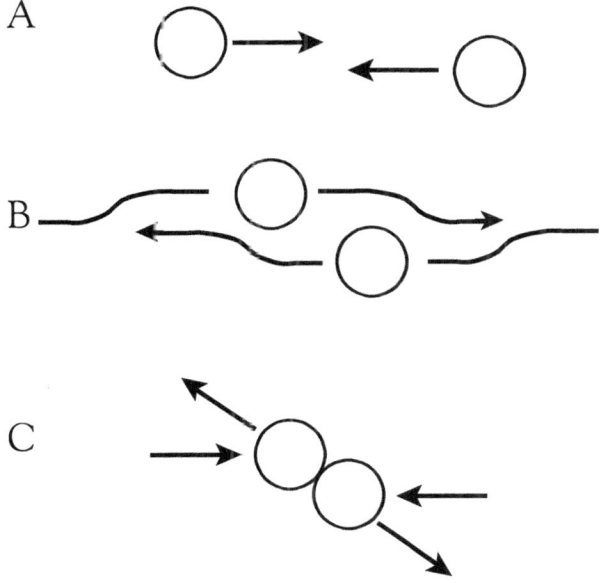

Figure 8.2 Interactions Between Two Particles

and friction. Generally speaking, however, collisions are of the billiard ball type and the character of the momentum transferred during collisions depends on the elastic properties of the particle materials.

The numbers and intensities of collisions, of course, will depend on the number density of particles in suspension and the flow velocities. When particle suspensions are crowded, particles can bounce from collision to collision because there is no free volume within the suspension for them to move into to avoid collisions. When particles are not crowded in suspensions, collisions may be few and far between.

Particles Moving in the Same Direction

The particles in Figure 8.2 are moving in opposite directions. You may be wondering if they won't normally be moving in the same direction in most flow circumstances? Yes, normally, they will be traveling in the same general direction. The shear conditions that must be considered, however, are the imposed shear rates. Even when all particles are moving in the same general direction, the higher the shear rate, the higher the velocity differences can be between adjacent particles.

When one particle is moving at 100 cm/sec and another is moving at 150 cm/sec, the difference between the two (the approach velocity) is nevertheless 50 cm/sec. Taken from the viewpoint of either particle, the other particle is still approaching at 50 cm/sec.

The simplest picture to make this case is to consider a drive on a German Autobahn. When you are driving down the Autobahn at 160kph in your rental car and a _____ (fill in the blank with your favorite European sports car) comes barreling past at over 240kph, you would think you were standing still. Even though you're going really fast, it doesn't feel like it relative to the sports car. That's the point! And if the two cars had a collision, it would be a whale of a collision! (... even though both cars were headed in the same direction.)

In pipe flow at high shear rates, a few micrometers distance can represent an enormous velocity difference due to the large velocity gradient known as the *shear rate*.

Summary

The carrier fluids in suspensions will flow around obstacles. When there is sufficient fluid momentum to influence and accelerate entrained particles out of the path of obstacles, entrained particles will flow around the obstacles as well.

When particle momenta are too great for fluid momentum to sufficiently alter their trajectories, particles will collide with obstacles. When suspension solids contents are high so the particles are crowded together, collisions will be frequent and dominant.

Collisions can occur between particles in any particulate/fluid suspension. The numbers and intensities of collisions will depend on the number density of particles per volume of suspension, and the flow velocities of the suspended particles.

Colloidal particles are small enough so they will travel with carrier fluids and rarely collide with other particles. But when their velocities are sufficiently high or they are contained in high solids content suspensions, they too are subject to collisions. Coarse particles will frequently collide with any obstacles in their paths. When their velocities are sufficiently low, and they are in low solids content suspensions, coarse particles can flow smoothly around obstacles.

All particles with sizes between these two extremes are similarly subject to collisions. The two main properties that govern such interactions are: particle/fluid flow velocities and suspension solids contents. When flow velocities and solids contents are high, collisions can be expected to dominate suspension flow. When flow velocities and solids contents are low, collisions can be rare.

Chapter Nine

Repulsive Forces and Deflocculation

Deflocculation causes interparticle repulsive forces to dominate within particle/fluid suspensions. When suspensions are deflocculated, shear-thickening rheologies, i.e., dilatant behavior, can be expected to appear. The interrelationships between these phenomena will be covered in this chapter.

Repulsive Interparticle Forces

pH Effects

When particles suspended in fluids exhibit highly positive or highly negative electrostatic surface charges, repulsive forces will dominate and the particles will repel each other. Under such conditions, all particles will remain separate and as far distant from one another as possible.

The mineralogical type of the particles, the pH in the fluid environment, and the presence and concentration of deflocculating additives all affect the sign and charge densities of electrostatic repulsive surface charges.

As mentioned earlier, each different mineralogical type will have an *isoelectric point* (IEP). The isoelectric point is the pH at which the net surface charge on the mineral is zero. Under pH conditions that are more acidic than the IEP, i.e., at lower pH values, electrostatic surface charges will be positive. Under pH conditions that are more basic than the IEP, i.e., at higher pH values, electrostatic surface charges will be negative.

In clays and other minerals where plate surfaces and edges are not necessarily the same charge, these tendencies still generally hold relative to the effective IEP values for the surfaces and for the edges.

At the IEP, suspensions will flocculate. As pHs move away from the IEP, suspensions will begin to deflocculate. At pHs far from the IEP, suspensions will be highly deflocculated.

In deflocculated suspensions, the weak van der Waals attractive forces are masked by the stronger electrostatic repulsive forces. Van der Waals forces **are** present at all times (they can't be eliminated), but they are weak and easily overpowered by the stronger electrostatic repulsive forces.

Deflocculant Effects

Anionic Polyelectrolytes

Chemical additives that enhance the electrostatic repulsive forces are called *deflocculants*. These frequently take the form of anionic polyelectrolytes, many of which are long-chain hydrocarbons (polymers) with many ionizable entities along the length of the polymer chains.

When anionic polyelectrolytes are added to particulate suspensions, cations located periodically along the polymer chains ionize and are released into solution. After the positively charged entities are released, the many negative sites along the polymer chains cause the polymers to be highly negatively charged.

When such polymers are added to suspensions, they coat particle surfaces just as paint coats the surfaces of a wall. Even if the particles are already electrostatically negative, the coatings can enhance the particles' negative surface charges. The suspended particles will then become more highly negatively charged, and the suspensions will be more highly deflocculated.

Because long chain polymers are hydrophobic ("water hating"), the water minimizes its contact with them by pushing them out of the suspension. Since the polymers can't totally be removed from suspension, the water will push them onto any available surfaces where only one side of the polymer remains in contact with the water. Thus, contact between the water and the polymers is minimized.

Hydrophobic forces are sufficiently strong to overcome electrostatic repulsive forces. For this reason, anionic organic polyelectrolytes function well when added to suspensions containing positively or negatively charged particles. The hydrophobic effect pushes the electrostatically negative polymers out of the carrier fluid and onto the particle surfaces. Even though one would expect negatively charged polymers and negatively charged particles to repel, hydrophobic forces are stronger than electrostatic forces and the polymers will coat the particles anyway, enhancing their negative surface charges.

When deflocculants are added to suspensions containing positively charged particles, the negatively charged polymers can cancel and mask positive surface charges, and then impart negative surface charges on the particles. Depending on the types of particles and the pH of the suspension, relatively low concentrations of some deflocculants can change positive surfaces to highly charged negative surfaces.

Another aid to deflocculation provided by the organic polyelectrolytes is known as *steric hindrance*. The word *steric* defines this as a spatial effect. As the polyelectrolytes lay on the particle surfaces, they are similar to a layer of paint coating the particles. When two such particles closely approach one another, the actual particles cannot come into contact because they are prevented from doing so by the layers of additive coating each surface.

When organic polyelectrolytes are used as deflocculants, not only do they enhance particle surface charges so the particles repel one another, but they also hinder particles from coming any closer to one another than the thickness of the additive layers on the approaching particles. Their coverage and effectiveness, of course, depends on their concentrations in the suspensions. Both mechanisms are characteristic of polyelectrolytes.

Inorganic Deflocculants

Another common deflocculant used in ceramic suspensions is sodium silicate, which is an inorganic additive. Sodium silicate helps to enhance the interparticle repulsive forces in suspensions by a fundamentally different mechanism than that of the many organic additives.

Sodium silicate is soluble, not hydrophobic. It enhances the negative surface charges in suspensions, not by adding more negative surface charges as the organics, but by removing positive flocculating cations. Mg^{++}, Ca^{++}, and Al^{+++} ions are examples of common multivalent flocculating cations. In solution they are attracted to the negative surface charges and they effectively cancel those charges and reduce the interparticle repulsive forces. Magnesium, calcium, and aluminum silicates are all insoluble, however. So when sodium silicate is added to suspensions, the silicate ions can combine with these multivalent cations and precipitate in the form of insoluble silicates.

As long as the soluble cations are in solution, they can be loosely associated with the particles' negative surface sites. They can be attracted to a site, loosely combine with it and cancel its charge for a fraction of a second, and then they can move back off into the interparticle fluid. The soluble cations are dynamically associated with the negative surface sites. The stronger their charge, the longer they can stay associated with one or more negative surface sites. But as ions in solution, they move about, back and forth, nearer and then farther away from the particle surfaces.

When these multivalent flocculating cations combine with a silicate anion, they are precipitated, neutralized, and removed from solution. As cations, they move dynamically about within suspensions, but once precipitated as insoluble silicates, they no longer function as flocculating cations.

In this way, sodium silicate also enhances the negative particle surface charges in suspensions by removing flocculating cations. It follows a different mechanism than the organic polyelectrolytes, but the end result is similar.

Interactions Between Deflocculated Particles

Suspended particles that are highly charged electrostatically (either positive or negative) will repel one another. Such particles will not easily come together to form gel structures. Figure 9.1(A) is the same suspension snapshot shown in Figure 6.3(A). Figures 9.1(B), (C), and (D) depict suspension snapshots taken at later times when the system shown in Figure 9.1(A) is maintained under deflocculated conditions.

As deflocculated particles move around in suspensions, they try to stay as far from one another as possible due to repulsive interparticle forces. Figure 9.1 depicts a time period similar to that in Figure 6.3. Particles are free to move around, but flocculation does not occur and gel structures do not form when repulsive electrostatic forces dominate as they do in deflocculated systems.

Suspension Instability

Due to the lack of gel structure in deflocculated systems, as depicted in Figure 9.1, it is not only possible, but likely, that large particles will settle. Whether or not this occurs depends on the solids content of the suspension and the size of the coarsest particles. Most colloidal particles are small enough that they will not settle appreciably, but coarser particles can and will settle when conditions permit. Typical ceramic suspensions consist of many particles in each of these categories.

Particle settling rates will be by order of size. The coarsest particles will settle the fastest, followed by the next coarsest, etc. When solids contents are sufficiently low, smaller particles can shift out of the way to allow larger particles to settle. Extremely low solids contents provide appropriate conditions for unhindered settling to occur. Such conditions (unhindered settling) are required for particle size analysis by sedimentation.

Settling velocities, which can be calculated using Stoke's Law (Equation 3-2), are functions of the mass of each particle and of the viscosity of the carrier fluid. The masses of colloidal particles,

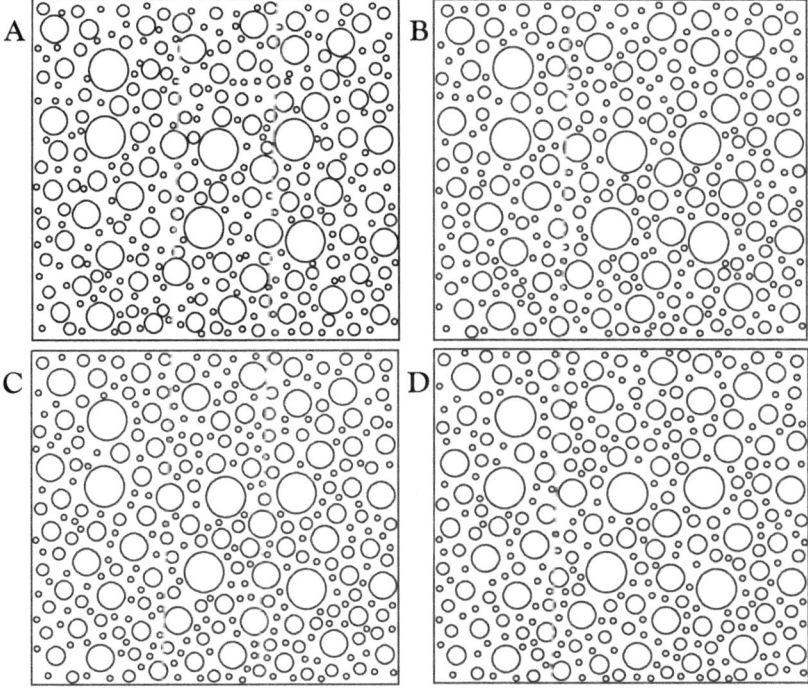

Figure 9.1 A Deflocculated Suspension

however, are small enough that they are affected by Brownian motion. Brownian motion of colloids can be in any direction – with or against gravity. Colloids therefore do not settle with predictable velocities due to Brownian motion. They tend to move around in random directions. Colloidal particles are also small enough to be greatly influenced by the repulsive interparticle forces in deflocculated suspensions and by the flow of carrier fluids.

Conditions required for particle size analysis by sedimentation are extremely low solids contents and highly deflocculated suspensions. Such sedimentation techniques are useful to analyze particle sizes down to, but not including, the colloids.

The tendency, therefore, is for all non-colloidal particles in deflocculated suspensions to settle. If solids contents are low enough, and deflocculation levels are strong enough to allow particles to move around locally without coming too close to other particles, coarse particles can settle to the bottoms of containers.

If samples of deflocculated and flocculated suspensions are placed in glass containers and allowed to settle, the results will be fundamentally different. Although the flocculated suspensions will form gel structures, settling can still occur depending on the solids content of the suspensions. When flocculated suspensions settle, all particles, including the colloids, will be tied into the gel structure, and the liquid at the top of the container will be clear. A flocculated system will have a well-defined sediment/supernatant fluid interface, and the supernatant fluid will be perfectly clear.

When a deflocculated suspension is allowed to settle, the coarse particles may settle to the bottom of the container, but there will be no clear sediment/supernatant fluid interface. Fluid near the suspension surface will be hazy because colloids will be moving around in that fluid. Colloids will not settle, nor will they be pulled into a gel structure. They will remain separate from the other particles. Depending on the suspension conditions, deflocculated suspensions may appear unchanged, even after sitting for 24 hours on a shelf.

The key factor that distinguishes **deflocculated** suspensions from **flocculated** suspensions is that particles in deflocculated suspensions travel independently of one another. In flocculated suspensions, particles travel in groups (flocs) of various sizes.

One does not usually expect settling to occur in flocculated systems, because all particles should quickly be incorporated into the growing gel structures. When agitation levels are sufficient to break the gel structure, solids contents are low enough, and coarse particles are large enough, it is nevertheless possible for coarse particles to settle, even in flocculated suspensions.

When storage tank agitators are properly designed to produce vertical recirculation, settling should not be a problem.

Shear-Thickening Rheologies

Flocculation and gel structures tend to cushion the interactions between particles, but deflocculation minimizes cushioning effects and allows particles to collide more easily. As shear rates increase, and particle-particle collision rates increase, apparent viscosities can increase as well.

A computer model, written by one of the author's former students[3], simulated particulate suspensions during flow. The goal of the model was to calculate apparent viscosities at various levels of shear. In this model, the only form of energy transfer allowed was transfer due to particle collisions. Attractive and repulsive interparticle forces and gravitational forces were ignored. The upper surface of the cell moved at various velocities, while the lower surface was stationary. Only particle/particle collisions and collisions between the particles and the upper and lower surfaces could transfer energy within the model cell.

The results of this model showed **only** dilatancy. As the velocity of the cell's upper surface increased, calculated apparent viscosities increased. As the solids concentration of particles in the cell increased, apparent viscosities increased. The results of this model showed very clearly that particle-particle collisions produce shear-thickening (dilatant) rheologies.

Other models written at about that same time covered conditions at the other extreme where solids contents and shear rates were very low. Those models included algorithms to simulate the interactions due to attractive and repulsive interparticle forces. Algorithms to handle collisions were not included in those models, so those particles could not collide.

The results of those models demonstrated that interparticle forces caused shear-thinning behaviors. They showed **only** shear-thinning behaviors. They did not demonstrate dilatancy because collisions were not allowed. We didn't see shear-thinning behavior in our model's results because we ignored interparticle forces. All of the results fit together very nicely.

Particles traveling individually and independently in deflocculated suspensions will show dilatancy, because when the particles interact, those interactions will include particle collisions. As shear rates increase, the number and intensities of collisions will increase, and the measured apparent viscosities will increase as well.

Dilatant behavior will also occur in flocculated suspensions, but it will usually only be readily apparent at high shear rates where all remnants of gel structure have been destroyed and particles are traveling as individuals. As shear rates then continue to increase, particle-particle collisions will again increase in number and intensity and apparent viscosities will increase.

Summary

Deflocculated systems are characterized by particle/particle repulsion, relatively low viscosities, settling, the absence of gel structure, and dilatancy. When fluidity is required in extremely high solids content suspensions, high levels of deflocculation are required. Under such conditions, all of these properties can be expected.

Chapter Ten

Dilatancy

When particle/particle collisions dominate during suspension flow, dilatancy occurs. Shear-thinning behavior results from the breakdown of gel structures as suspensions are sheared. Dilatancy, however, occurs when particles collide during shear.

Unlike shear-thinning behaviors which usually are accompanied by yield stresses, dilatant rheologies with and without yield stresses are common.

The Physical Explanation for Dilatancy

As solids contents increase, particles are forced closer and closer to one another. As high solids suspensions and forming bodies are sheared, particles interact frequently. Whether the interactions grow from being close encounters, to glancing blows, to major collisions depends upon both the suspension properties and the applied shear rates. Close encounters and glancing blows between neighboring particles in suspensions are normal and to be expected. Intense particle/particle collisions, when solids contents and/or applied shear rates are high, can be detrimental to the viscous properties of suspensions.

Consider Figure 10.1 which shows arrangements of close-packed spheres before, during, and after a shear stress has been applied and the particles have rearranged. Figure 10.1A shows the initial arrangement of packed spheres. Figure 10.1B shows the upper layer as it is moving up and over the lower layer. Figure 10.1C shows the final arrangement after the upper layer has been displaced one sphere to the right.

The packing density of dense-packed 3-D arrangements (hexagonal close packing) of spheres, depicted in Figures 10.1A and 10.1C, is 74.04%. Simple cubic arrangements, depicted by Figure 10.1B, have packing densities

Dilatancy

Packing Factors

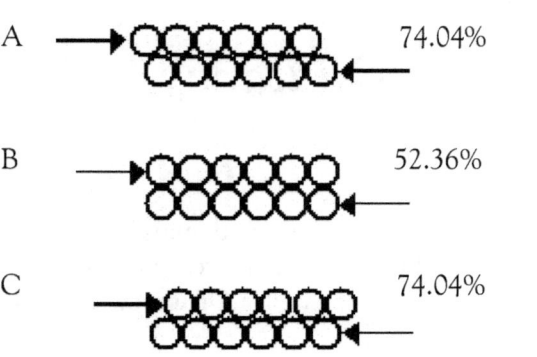

A   74.04%

B   52.36%

C   74.04%

Figure 10.1 Schematic of Dilatant Reaction to Shear

of 52.36%. Dilatancy exists when particles pass through less dense arrangements as they are sheared. When this occurs, particles move closer together and they frequently come in contact with one another, opening the structure.

In Figure 10.1, this occurs as the densely packed spheres in (A) pass through a less dense arrangement (B) as they are sheared to rearrange to form (C). The structure of Figure 10.1B has opened, that is, it has *dilated*, relative to both Figures 10.1A and 10.1C. When shear is removed, the particles can again settle back into an arrangement similar to their original arrangement, with its higher packing density, in Figure 10.1C.

When a highly dense system of particles in suspension is sheared, collisions between particles cause local structures to open, to *dilate*, during shear — hence, the name *dilatancy*.

In process suspensions, particles will not be in contact with one another as shown in Figure 10.1. Only after ware are formed and the systems are dry (or drying) would one expect all particles to be in contact with one another. In suspensions, the particles depicted in Figures 10.1A and 10.1C should be separated by fluid, but they may still go through local rearrangements during shear in which the particles will touch as in Figure 10.1B.

When shear rates are relatively low, and particles are separated by interparticle fluid (that is, when each particle has some space to move around without contacting any other particle), it is possible that they can move past one another without colliding and without causing the structure to dilate. But as shear rates increase, more and more collisions occur and dilatant behavior appears.

To demonstrate this, consider a slightly different set of circumstances than the previous example. A suspension contains 60 volume % particles and 40 volume % fluid. Particles are packed in a dense packing arrangement as in Figure 10.1A, but some fluid separates all particles. That is, no particles are touching. Since a dense packed arrangement of spheres packs to 74.04% (i.e., 25.96% porosity), 40 volume % fluid is more than enough fluid to fill all 25.96% of the pore volume and excess fluid remains. The 14.04 volume % of excess fluid can then separate all particles and put some distance between them.

In a perfect arrangement, all particles in this example at 60 volume % solids will be separated by fluid, and no particles will be in contact. When this suspension is sheared at low shear rates, particles can move past one another without colliding. Some particles will move closer together so other particles can freely change positions during shear. At higher shear rates, however, particles will continue to rearrange locally, but other particles will begin to collide. Then, as higher shear rates continue, more and more particles will collide and large numbers of particles will be forced into less densely packed arrangements like that in Figure 10.1B.

The simple cubic arrangement of Figure 10.1B contains 47.64% pores, which is more than the 40 volume % fluid in the example. If shear forces particles in this suspension into a simple cubic arrangement, it will no longer have sufficient fluid to fill all pores. Depending on the magnitude of the driving force that is producing the shear rates, large regions within the 'suspension' may contain simple cubic arrangements of particles. Under such conditions, many particles in those regions will be forced to contact one another with no fluid surrounding them or filling adjacent pores. Such arrangements no longer fit the definition of a 'suspension.'

When this occurs, the system will behave more like a bunch of wet particles than like the suspension it is at low shear rates. Apparent viscosities will then be extremely high (off the scale) and when sheared at high rates with sufficient stress, the system of particles in the open (and relatively dry) structures can actually rip or break apart like a weak solid rather than flow as a suspension. This is dilatancy.

The Onset of Dilatancy

When dilatancy, or the possibility for dilatancy, exists in process suspensions, one must pay close attention to the *onset of dilatancy*. This is the shear rate at which dilatant properties begin.

Figure 10.2 shows a viscosity versus shear rate rheogram in which the *onset of dilatancy* is marked. The 'onset' of dilatancy

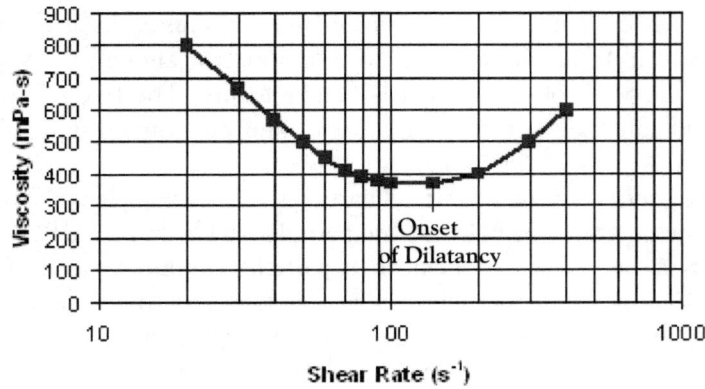

Figure 10.2 Onset of Dilatancy

occurs at the shear rate at which the apparent viscosity begins to increase. In Figure 10.2, the onset of dilatancy occurs at about $150s^{-1}$. Notice in this figure that at shear rates below the 'onset', the rheogram shows a shear-thinning suspension, and at shear rates above this value, the rheogram shows a dilatant suspension.

There are two important aspects to the onset of dilatancy. First, *one must be familiar with process suspensions* to know where each begins to show dilatancy. That is, one must measure the rheograms of process suspensions to characterize the shear rates at which the onset of dilatancy in each occurs. Second, *one must be familiar with the process itself* to know the magnitude of shear that will be imposed upon the suspensions at each step of the processing.

At what shear rate does each suspension begin to exhibit dilatancy? What is the highest shear rate applied to each suspension and exactly where in the process does it occur? These answers need to be known. With these answers in hand, one can predetermine whether dilatancy will be a problem in a process system.

Many rheograms simply show shear-thinning behavior with no hint of dilatancy as shown in Figure 10.3. When the apparent viscosity approaches

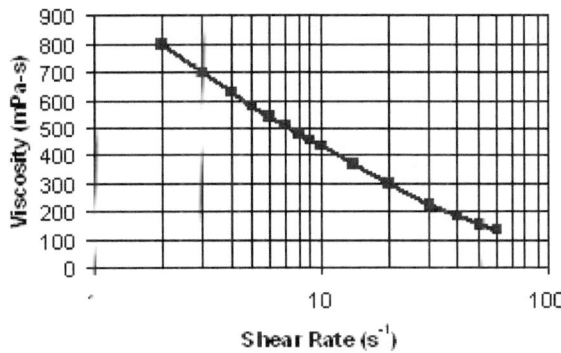

Figure 10.3 Typical Yield Shear-Thinning Rheogram

low values as in this figure, the minimum viscosity must be near. If the rheogram in Figure 10.3 covers the full range of typical process shear rates, however, dilatancy should not be a problem for this suspension in this process. If the rheogram in Figure 10.3 does not cover the full range of process shear rates to which this suspension will be exposed, then dilatancy may be a problem when process shear rates exceed the maximum measured value, which in this case is ~60s⁻¹.

Many common rotational viscometers are limited to relatively low shear rates. If their rheograms show any dilatant behavior at all, dilatancy can be expected to be a problem during processing.

Where Must One Watch for Dilatancy?

When suspensions are dilatant, process shear rates must remain low. It is preferable that suspensions never enter the dilatant regime, which occurs when shear rates exceed the shear rate at the onset of dilatancy. The onset of dilatancy, however, is a function of the solids content, the particle size distribution of the powder, and the type and concentration of additive chemicals. It can be adjusted.

The higher the solids content, the higher the concentration of deflocculant additives, and the higher the applied shear rates, the more likely it is that a system will enter the dilatant regime and cause processing problems.

Two of these three factors, the solids content and the concentration of additives, are suspension factors. If a suspension has been tuned and is sent to the processing floor, the only factor remaining to adjust is the applied shear rate. If the suspension on the process floor is dilatant, under such conditions, one must make sure that applied shear rates remain **low**.

Where are the high shear rate processes to which dilatant suspensions will be subjected? Pumps, forming operations, and some finishing operations can produce high shear conditions. Atomization is an extremely high shear operation. Injection molding can be a high shear operation. Pipe flow and extrusion are usually low to medium shear rate operations, but flow through pipe nozzles and extrusion dies can be high shear.

When suspensions are dilatant, one **must** pay attention to the results produced by the high shear operations. When necessary, the processing rates in high shear operations must be reduced so they become low shear operations. If that is not possible, the suspensions or forming bodies must be adjusted to reduce their dilatant character.

Can Dilatancy Be Eliminated?

Since dilatancy is the result of particle/particle collisions, all systems that contain flowing particles can be dilatant if they're sheared fast enough to cause the particles to collide. It appears that all suspensions will enter a region of dilatancy if they are sheared at high enough shear rates. To completely *eliminate* dilatancy, one must eliminate all particle/particle collisions. The best way to accomplish this is to remove all particles. In ceramic process systems, this is obviously not possible.

Fortunately, most extremely high solids systems are not operated at high shear rates (extrusion, for example) and most extremely high shear rate operations (such as atomization in spray dryers) are operated at low viscosities using low solids content systems. In both of these examples, dilatancy could be a major problem, but it is usually not a problem.

Some processes are operated at extremely high solids contents, and others at extremely high shear rates. Forming bodies and suspensions used in both such processes can easily enter the dilatant regime through random, day-to-day, particle size distribution fluctuations. This doesn't happen often, but it can and does happen.

To *minimize* dilatant effects (notice this doesn't say "to *eliminate* dilatant effects"), one must consider how to reduce particle/particle interactions. Some obvious answers are to lower solids contents, to lower shear rates, and/or to increase the level of flocculation of the suspensions. Since the addition of flocculants causes apparent viscosities to rise, it may be necessary to lower solids contents at the same time as flocculants are added.

For a variety of reasons, many ceramists attempt to set solids contents at their maximum possible values, which invariably also requires the additions of deflocculants. Both of these adjustments (higher solids and higher deflocculant concentrations) decreases the shear rate at the onset of dilatancy.

Consider the rheograms shown in Figure 10.4. When suspension A is deflocculated to produce suspension B, or diluted to

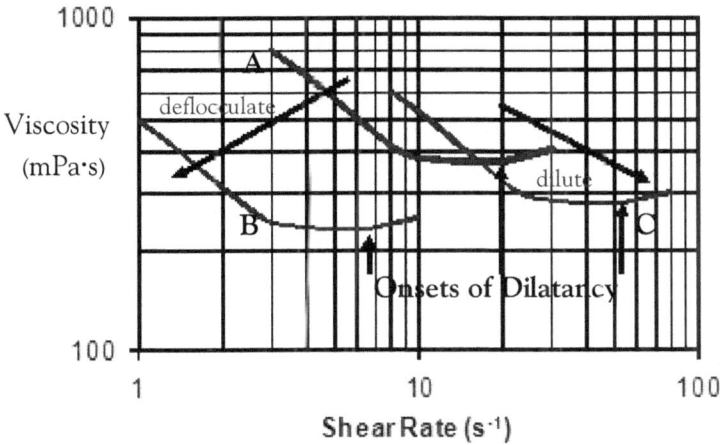

Figure 10.4 Effects of Dilution and Deflocculation

to produce suspension C, the resulting viscosities of both suspensions decrease. But deflocculation brings the onset of dilatancy to lower shear rates and dilution pushes the onset of dilatancy to higher shear rates.

The opposite counterpart of each of these is also demonstrated by this figure. When suspension B is flocculated to achieve suspension A (which proceeds in the opposite direction of the *deflocculate* arrow), the viscosity not only increases, but the onset of dilatancy moves to higher shear rates. And when the solids content of suspension C is increased to achieve suspension A (which proceeds in the opposite direction of the *dilute* arrow), the viscosity also increases but the onset of dilatancy moves to lower shear rates.

Solids content and chemistries may each be adjusted alone, but they are frequently adjusted together. If a suspension's apparent viscosity is good, but the *solids content is too high*, the suspension will typically be diluted (thereby reducing apparent viscosity) to adjust solids content to its target value, and flocculated to return the suspension to its original apparent viscosity. Or if the suspension's apparent viscosity is good, but the *solids content is too low*, the solids content can be increased to its desired level (thereby raising the apparent viscosity) and deflocculant can be added to bring the apparent viscosity back down to the target value.

Note in these examples that each *pair* of adjustments moves the onset of dilatancy in the same direction. In the first case, *diluting and flocculating* both move the onset of dilatancy to higher shear rates. In the second case, *raising* the solids content *and deflocculating* both move the onset of dilatancy to lower shear rates.

Solids content and additive chemical adjustments are both excellent ways to change the apparent viscosities of suspensions. When one uses these adjustment techniques, however, one must be aware that in addition to apparent viscosity changes, they each affect the shear rate at which dilatant properties appear. When used together (as in these examples) to cancel each other's effects on apparent viscosity, they don't cancel each other's effect on the onset of dilatancy.

Can Dilatancy Be Beneficial?

The author has met individuals who believed dilatancy in their forming bodies was beneficial to their processes. In the author's opinion, very few such processes, if any, ever benefitted from the use of dilatant forming bodies.

The only places *mild* dilatancy appears to be beneficial is during milling and high intensity mixing operations. Note that the adjective used to describe such dilatancy was **not** *extreme*, but *mild*.

High intensity dispersion (HID) systems and milling operations are designed to deagglomerate and to mill systems of agglomerates and large particles. In such systems, *mild* dilatancy can enhance the particle/particle collisions which can enhance the milling and deagglomeration operations.

Our experience with high intensity dispersion suggests that it only works properly at sufficiently high solids contents where particle collisions are frequent. The same is true of milling operations. Low solids contents in either operation makes it unlikely that the particles will be impacted by the dispersion blade or by the milling media. Extreme dilatancy in either case will be detrimental. Viscosities will be too high, so mixing and milling operations will be hindered.

The author once saw extremely dilatant bodies being mixed in a lab in a common, backyard cement mixer. These bodies were so dilatant that under the influence of gravity on a tilted surface, they flowed down the surface only very slowly. As the mixing chamber rotated, the ball of body in the bottom of the slightly tilted chamber was flowing so slowly due to the dilatancy that only its outer surfaces ever came in contact with the mixing blades. Before the blades could cut into the body and actually do some mixing, the rotation of the mixing chamber carried the body to the higher elevation of its circuit and the body flowed slowly away from the blades.

The recipe called for 30 minutes of mixing, so these bodies were rotated in the bottom of the mixer and their surfaces were dented by the mixing blades for 30 minutes but no mixing ever actually occurred. These bodies were no better mixed after 30 minutes than they were when they were initially placed into the mixers. When the author suggested that the mixer rpm needed to be decreased, he was told that the mixers only ran at one speed. The rpm of these mixers could, however, have been reduced by changing the pulley diameters or the motor rpm.

It is doubtful that the results of these experiments ever made any sense because the levels of mixing achieved were inversely proportional to the magnitude of the dilatancy exhibited by the test bodies. Since all of these bodies were extremely dilatant, hardly any mixing occurred in any of them. All tests in this series were mixed for exactly the same length of time, so it was assumed that each of these samples was well-mixed to the same, constant level. The actual levels of mixing achieved depended on the natures and durations of the procedures employed before the samples were placed into the 'mixer' but the natures and durations of those procedures were not being quantified. The extreme dilatancy in this series of tests was certainly not beneficial.

Summary

Dilatancy results from particle/particle collisions within flowing suspensions. The only way to reduce dilatancy when it occurs is to make changes that reduce the magnitude and number of collisions. Shear rates can be reduced; particle size distributions can be changed; moisture contents can be increased; and the level of flocculation of the suspensions can be increased. All of these will help to reduce the dilatant properties.

When dilatant properties appear in bodies in the batch house (well before they must be sent to the process floor), suspensions can be altered and adjusted to reduce the dilatant properties.

However, when dilatant suspensions have been sent to the processing floor where they must be used, the only solution is to reduce process shear rates. In other words: **SLOW DOWN!**

Chapter Eleven

Syneresis

There are two phenomena that can cause major problems within ceramic process systems. Syneresis, which occurs at extremely high levels of flocculation, is one. Extreme dilatancy, which produces blockages at high shear rates and/or at high levels of deflocculation, is the other. Syneresis is a chemical problem. Dilatancy is a physical problem. Dilatancy was covered in the previous chapter. Syneresis will be covered in this chapter. Dilatant blockages will be covered in the next chapter.

Syneresis in a Suspension

Syneresis is the extreme densification of a gel structure accompanied by the expulsion of fluid from within the structure. It occurs when bodies or suspensions are in states of severe overflocculation. When a body is flocculated, a gel structure forms, such as the one shown in Figure 11.1.

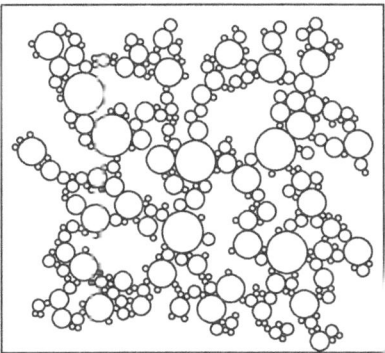

Figure 11.1 A Normal, Flocculated Gel Structure

Syneresis occurs when attractive forces within the gel structure are strong. After the gel structure has completely formed, as shown in Figure 11.1, the strong attractive forces continue to densify the structure and expel fluid.

When syneresis occurs in a suspension, the gel structure forms quickly, and as it densifies, the gel appears to settle, clear supernatant fluid is expelled from the structure, cracks are formed throughout the structure, and more fluid is expelled into the cracks and supernatant.

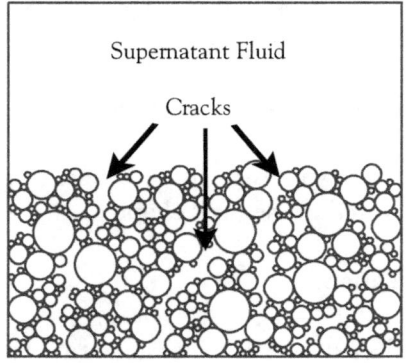

Figure 11.2 A Syneretic Structure

Figure 11.2 shows the structure from Figure 11.1 after it has been densified by syneresis. Figure 11.2 shows that the structure has settled to the bottom of the container, densified, expelled supernatant, and developed long channels (filled with clear fluid) that appear as surface cracks on the dense gel. Cracks that appear on the surface usually extend well into the structure. Other cracks can form within the structure that do not extend to the surface. Cracks in a syneretic gel are characteristic of syneresis.

Syneresis in a Plastic Forming Body

Syneresis is not limited to suspensions, but it also occurs in plastic forming bodies. The explanations and the figures are the same. The only difference is that plastic forming bodies start at higher solids contents than suspensions, and therefore, are even denser structures than those found in syneretic suspensions.

When plastic forming bodies are syneretic, the tell-tale sign will again be the formation of cracks throughout the body. There are quite a few reasons why cracks can form in a ceramic forming body. Syneresis is just one of them. But when cracks form in a quiescent, green body before drying has begun, syneresis is the probable cause.

Improperly adjusted bodies can rip due to dilatancy during forming operations. But when syneresis occurs, forming properties will usually be quite good. Flocculated bodies (even over-flocculated bodies) will typically be shear-thinning and they will shear and flow very nicely during forming operations. After the wares are formed and are no longer being sheared or handled, syneresis will then cause cracks to appear.

For example, after a filter pressing operation in a whiteware company, the filter cakes were stacked onto a cart to be moved to the next station. Large cracks formed in the stack of cakes and large chunks of the cakes fell off the sides of the cart. Syneresis was the cause.

Later, these same filter cakes were put through a vacuum extruder. The large cylindrical extruded columns were flowing nicely out of the extrusion die, but as they moved away from the die, cracks were again (still) forming. The cracks continued to form and expand even after the extruded columns were cut to length and the blanks were stacked on a pallet.

If attempts are made to plastically form such bodies, the ware can be expected to form well .... and then cracks will appear.

When syneresis occurs, bodies must be adjusted chemically to less flocculated states. The short term solution of course, when it is possible, is to try to fix the ware cracks. The long term solution, however, must be to adjust the body.

Evidence of Syneresis

A simple gelation test with a recording viscometer can show evidence of syneresis. Figure 11.3 shows two twenty minute gelation curves typical of those that can be made with a rotational viscometer at constant shear rate. In this figure, the one gel curve exhibits syneresis, the other gel curve does not.

The suspension that does **not** exhibit syneresis shows an excellent gelation behavior as the apparent viscosity increases quickly (initially), and then increases more slowly with time towards the final limiting viscosity. The syneretic suspension also shows excellent gelation behavior as the apparent

Syneresis

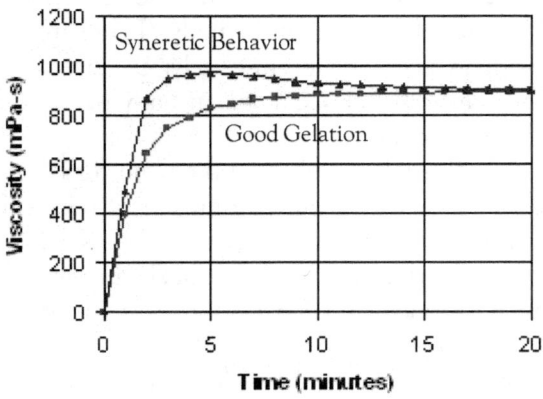

Figure 11.3 Good versus Syneretic Gelation Behavior

viscosity quickly increases towards the final viscosity. But the viscosity of the syneretic suspension quickly exceeds the final limiting viscosity and then, with time and constant shear conditions, the viscosity decreases slowly toward the limiting values.

Note the difference between the two cases. Syneresis produces apparent viscosities at low measurement shear rates that quickly exceed and then **decrease** with time toward the limiting value. Non-syneretic, flocculated suspensions exhibit apparent viscosities at low measurement shear rates that **increase** with time toward the limiting values as the gelation processes build structure.

Excellent gel properties are in evidence when apparent viscosities initially increase quickly towards the limiting values and then continue to increase more slowly to approach the limit with time. Gel properties are also excellent, but too strong and syneretic when apparent viscosities quickly exceed the limiting value and then fall back to it with time.

The Cause

Extreme levels of flocculation, whether intentional or not, cause syneresis. The presence of high concentrations of flocculating cations, such as Ca^{++}, Mg^{++}, and Al^{+++}, can produce syneretic conditions.

High concentrations of flocculating ions can be the result of routine slip adjustments by process engineers. High concentrations of such ions can also be due to the choice of minerals used in a body, the use of hard water, and/or the strict adherence to a fixed body formulation.

Strict Adherence to a Body Recipe

Strict adherence to a body recipe is usually based on the assumption that mineral properties do not vary from day to day. When mineral suppliers ship perfectly consistent materials with perfectly constant properties, fixed body recipes **will** produce constant daily body properties. But mineral properties, no matter how tight the controls, always vary slightly, so constant body recipes will always produce bodies with properties that fluctuate from day to day.

Preparation of batches from day to day that always strictly adhere to the same body recipe can allow ion contents in batches to fluctuate over wide ranges. On the days when flocculating ions reach high concentrations, syneresis can occur.

Routine Slip Adjustments

An SOP (standard operating procedure), that calls for the addition of flocculants until a certain viscosity is achieved, can also produce syneresis. Particle physics and solids content variations during processing can occasionally produce slips with low viscosities. When excessive levels of flocculants are required to reach target process viscosities, syneresis can occur.

It is also possible that so many ions are present in the interparticle fluids that fluid conductivities are high and the effectiveness of additives is reduced. In such cases, excessive concentrations may be required to achieve (or just approach) desired body properties. Again, under these circumstances, syneresis can occur.

Choice of Minerals

Some minerals are insoluble in water; some are soluble; and some are partially soluble. The partially soluble minerals can cause syneresis as suspensions age.

For example, consider dolomite which is calcium-magnesium carbonate. Dolomite is very slightly soluble. With dolomite in a body, calcium and magnesium ions (flocculating ions) slowly dissolve and enter the carrier fluid. In suspensions, these flocculating ions tend to enter and be tied up by the gel structures. As they are tied up, more ions slowly continue to dissolve to replenish the supply in the carrier fluid.

Dolomite is the source for a potential overabundance of flocculating magnesium and calcium ions in suspensions and forming bodies. For the same reason, dolomite is a potential cause for syneresis.

Why is any particular mineral used in a body? There are many good reasons for using each mineral in ceramic production bodies. But since most ceramic bodies are aqueous suspensions, partially soluble minerals can be sources of troublesome ions. This should be considered when minerals are selected to be body ingredients.

The Solution

When syneresis occurs, the levels of flocculation must be reduced. Chelating agents and/or deflocculants should be added to adjust bodies or suspensions to less flocculated states. Each potential mineral source of the flocculating ions should be examined to determine if an alternative mineral can be used to solve or reduce the problem. When the slow dissolution of flocculating cations from certain minerals is identified as the problem, chelating agents may remove those ions, but aging can allow them to be replenished.

Syneresis does not respond to process changes. Since it is caused by improper chemistry, it needs to be remedied by adjustments to body chemistry. Some problems may disappear after making process changes – but syneresis is not one of them.

Summary

The important point to remember concerning **syneresis** is that it is a **chemical** problem. Since syneresis is a chemical problem, the long-range solution to syneresis **must** be a chemical adjustment.

Syneresis is produced by excessively high concentrations of flocculating ions and/or other flocculating chemical additives that cause bodies and suspensions to be strongly flocculated.

Flocculation and gelation occur quickly in syneretic suspensions. The concentrated additives then pull the gel structure into even denser arrangements. Clear, interparticle fluid is expelled from the structure as syneresis occurs.

Quiescent syneretic slips will appear to be settled, cracked, and covered with clear, supernatant fluids. Syneretic plastic forming bodies, will have excellent rheological properties. But after forming processes are complete, syneresis will further densify the body, expel carrier fluid, and produce the characteristic cracks.

Chapter Twelve

Dilatant Blockages

As we have discussed in earlier chapters, dilatancy is caused by particle/particle collisions. Under extreme circumstances, particles not only collide, but structures begin to form that can block flow channels. These types of structures are known as *dilatant blockages*. They are crowded compacts of particles that are mechanically bound together.

Dilatant blockages are not caused by, nor can they be solved by, chemistry. To be sure, high concentrations of deflocculants can exacerbate dilatant properties and increase the likelihood of dilatant blockages. But dilatant blockages are formed when particles are forced into mechanically linked structures. Mechanical dispersion forces alone can redisperse the particles in such blockages.

Syneresis, which was discussed in the previous chapter, is a chemistry problem. Dilatancy and dilatant blockages are mechanical problems.

The Formation of a Blockage

When particle/particle collisions become extremely intense, they can cause *dilatant blockages* to form. When extreme dilatancy characterizes a suspension, particle/particle structures can form and increase to the point where the dilatant structure spans the whole cross-sectional area of the flow channel. When this happens, flow stops. And viscosities are no longer measurable because dilatant blockages do not have viscosities.

A dilatant blockage is characterized by particles that are all mechanically locked in position against one another. When a dilatant blockage spans the complete cross-sectional area of a channel, flow ceases. When a dilatant blockage fills only part of a flow channel, flow decreases as the effective cross-sectional area of the channel decreases. When a complete dilatant blockage occurs, increased pressures and higher stresses behind the blockage further compact the particles and strengthen the blockage.

Some might think that removal of all shear stresses will allow such blockages to dissipate. Removing or relaxing the stress on a blockage, however, does **not** guarantee that it will break up and dissipate. In most cases, once a blockage has formed, it won't break up or allow particles to easily redisperse.

Mechanical dispersion is usually required to redisperse particles caught in dilatant blockages. This could occur when a channel is only partially blocked. Particles in the flow stream may impact particles on the outer surfaces of the blockage, dislodge them, and effectively redisperse them. For this to happen, however, the intense conditions that caused the blockage in the first place must have been eliminated. Local conditions must favor continuous suspension flow rather than the buildup of another blockage. Then, and only then, is it possible that flowing particles **MAY** break up and redisperse partial blockages. Chances of this happening, however, are slim to none.

One of two phenomena usually occurs when a dilatant blockage has formed: (1) Pressure builds up behind it and the blockage is pushed forward, accompanied by severe abrasion where its outer edges slide against the walls of the flow channel; or (2) all flow ceases, and pressures and stresses build up behind it until something breaks.

A Dilatant Blockage Is A Filter Pressed Cake

A simple picture of a dilatant blockage is to consider it as a filter pressed cake that occupies and blocks a process flow channel. Since filter cloths are not normally found at any point in a flow channel, filter cakes should not be present. Blockages that act like filter cloths can form, however, when particles interact and tangle with one another due to dilatancy. When a pileup of particles bridges across a flow channel, it then acts like a filter cloth placed in the path of the suspension flowing down the channel.

After a complete blockage has formed and the cake builds, fluid can be forced through the pores of the blockage and the solids content within the cake can rise well beyond the solids content of the suspension.

The structure of the initial dilatant blockage will typically be quite porous because dilatancy produces relatively open structures. But once a blockage has formed, shear rates behind the blockage decrease dramatically, and suspension can be filter pressed rather densely against the blockage. For this reason, a filter pressed cake is an excellent picture of a dilatant blockage. The blockage itself takes the place of the filter cloth, and all suspension to follow forms the cake. Depending how long it takes to detect the blockage, whole sections of pipe or flow channel can be filled with filter pressed cake.

With this simple picture of a dilatant blockage, one may ask: How easy is it to redisperse such a blockage/cake in a pipe? Well ... how easy is it to redisperse a genuine filter pressed cake without the use of a high speed agitator? Once a dilatant blockage has been formed, it simply will not redisperse easily, especially if it's located in a flow channel where there are no agitator blades to help. Without any flow, such as when the blockage is complete and the channel has filled with filter cake, it is effectively impossible to redisperse. When this occurs, take the clogged pipe out, throw it away, and replace it with a new section of pipe.

Two Examples

One example of a dilatant blockage occurred in a pelletizing operation where the goal was to pelletize what was essentially a high solids suspension of beach sand. The sand plus water formed an extremely dilatant body suspension. In this process, the suspension was to be extruded through a 3/4" thick hardened steel pelletizing die which contained many round channels to produce the pellets. The extruder was large enough to force this 'suspension' through the die plate. Blockages formed, but were pushed through the channels by the high extrusion pressures, abrading the channel walls as they slid along. The center of the die plate was chewed out by severe abrasion within about 45 minutes from the start of the test.

Fluid mechanics teaches that the fluid at the wall of a pipe remains stationary with the pipe. This may be true and accurate for a simple fluid, but it is not accurate when applied to suspension flow. The fluid in contact with the wall may remain stationary with the pipe (consistent with the theory), but particles in contact with the wall will not necessarily do so. In the pelletizing example, the particles at the wall were pushed along, scraping and abrading the hardened steel die until it was ruined — and that occurred fairly quickly.

With respect to particles in suspensions, Professor Funk from Alfred University, said many times over the years[8] that if the layer of particles adjacent to the walls of extrusion dies remained stationary in the dies, the dies would rust. But they don't rust – they are polished. This shows that the particles at the wall move relative to the wall.

Another example, which also came from Professor Funk, dealt with dilatant blockages in extrusion dies. He experienced a dilatant blockage in the die of a large production extruder. After the blockage formed, pressures built up until the die's mounting bolts (which were relatively large diameter bolts) broke and the pieces shot across the room like bullets fired from a rifle. According to him, you didn't want to be anywhere in the vicinity of the extruder when that happened.

Dilatant blockages can be dangerous as in this example. In other process environments, dilatant blockages may not be hazardous to humans, but they can easily damage delicate dies and cause major problems that hinder successful production.

Dealing with Dilatant Blockages

The immediate solution to the formation of a dilatant blockage is to stop the process and clean out the channel. Frequently, 'cleaning out the channel' (as suggested earlier) means to remove the blocked section of pipe or channel and replace it with a new piece of pipe or channel.

If the old channel is replaced with an identically sized new channel, the conditions will probably remain for another blockage to occur which can then ruin the new channel. If a dilatant blockage has occurred, it is an indication that the shear conditions in the channel are too great for the suspension being used. To reduce the shear conditions, the old blocked pipe can be replaced with a new one with a larger cross-sectional area. This will reduce the local velocities and shear rates in the flow channel; it will reduce the likelihood of another blockage forming; and it will reduce the chances that the new channel will be quickly ruined. Alternatively, the old pipe can be replaced with an identical new pipe, but then the flow rates and shear rates in that section of pipe should (must??) be reduced to prevent formation of another blockage.

Long range solutions to dilatant blockages are to modify the particle size distribution of the body, to add more fluid thereby reducing the body's solids content, to change the type and/or concentration of additives, or to do all of these in combination. All such modifications help prevent more blockages by reducing the frequency and intensity of particle/particle interactions which reduces a body's tendency to be extremely dilatant.

If a dilatant structure is forming, but it has not completely filled the flow channel, reducing the flow rate of the suspension in that section of pipe may cause the structure to dissipate, little by little, as the suspension flows slowly past and abrades the structure. This condition was discussed earlier. There is certainly no guarantee this modification will successfully remove a partial blockage, but it will definitely be slowing process flow rates in the mean while.

Detecting the formation of a blockage in a piping system or flow channel, however, is an extremely difficult challenge. In such systems, little evidence may appear until the channel is fully blocked. Blockages are more easily detected in extrusion or injection molding systems. Incompletely filled die cavities, or unwanted hollow channels in extruded columns will indicate the formation of such blockages.

After a dilatant blockage has grown and completely filled a flow channel, it is very unlikely that it will ever disintegrate. Pressure behind such a structure in a flow channel might cause it to slide along the channel until it either drops loose at an expansion of the channel (which would be lucky) or until it reaches a pipe fitting where it will become totally, mechanically locked into the pipe.

Once such a structure has formed, it is too late to make any corrections. In such cases, the only solution (the common solution, unfortunately) is to throw away the section of pipe that contains the blockage, and to start over.

When dilatant blockages occur within ceramic process suspension channels, they can be expected to be very strong because ceramic raw materials are strong in compression. The compression of a dilatant blockage, however, can place other process equipment into tension, so breakages may occur. For example, some injection molding dies contain delicate internal structures that can be easily broken when dilatant blockages occur. The initial blockage may simply prevent flow into downstream cavities, but upstream pressure on the blockage can build and eventually push the blockage with enough force to break through delicate die structures and ruin the internal structures of the injection molded parts. Such blockages may form again and again in the same location in each new die cavity. Body and/or process adjustments must be made in such situations.

A Simple, Non-Ceramic Example of a Dilatant Blockage

A fire drill provides the situation to demonstrate the formation and dissolution of a dilatant blockage. If a fire alarm sounds during a lecture class at a university and all of the students try to exit through the door of the lecture hall at the same time, a dilatant blockage will occur in the doorway. If twenty students are stuck in the doorway and the 21st student is pushing at the back of the pack, no one will be able to exit into the hallway. If the pressure by the students is sufficient, any student in the middle of the pack, and especially the two students at the front of the pack who are against the door posts, can be injured.

The 21st student at the back of the pileup really is in control of this situation. If he/she steps back, allowing all other students to then step back also, the blockage can dissipate and everyone can exit the room one at a time (from front to back) through the doorway. But if the 21st student continues to push against the back of the pack, no one will be able to exit the room.

To solve this problem, note that it is necessary that all students (from the back to the front) relax and take one step backwards away from the doorway. Then, one by one, from front to back, they can exit through the doorway.

The solution in this example is to relax the structure (everyone steps back) and to lower the flow rate (and the shear rate) by allowing one person to pass through the doorway at a time. This may be difficult to accomplish in an emergency situation such as in this example which takes place during a fire alarm. Everyone is excited, in a hurry, and wants to exit the room as quickly as possible, but the solution is to calmly exit the room slowly. It may be difficult to do under the circumstances, but it is the correct solution.

The same is frequently true in the process environment. When engineers are trying to run the process as quickly as possible, decreasing the speed of the process to prevent dilatant problems will be difficult to accomplish. Production process conditions and requirements hardly ever qualify as true 'emergencies' as in this example, but production goals may be driving process speeds up to the point where flow rates and shear rates are too high for the specific process and its process suspensions.

Also unlike the classroom fire drill example, when dilatant blockages occur in suspension flow, there is no easy mechanism that allows all particles to take one step backwards away from the blockage.

Particles locked into a dilatant structure will usually remain locked into the structure, even when stresses are removed. But stresses are not usually removed. When a blockage forms in a flow channel, pressures usually build up to the maximum possible output of the pump. When the operators notice that flow has ceased, then, and only then, are the pumps turned off and the pressures decreased to zero. But by then, it is too late.

Under the initial high pressure conditions following the formation of a blockage, the structures will densify and become more tightly packed. Axial pressure in a pipe will usually cause the structure to expand laterally and lock itself more tightly into the pipe. This happens for example in dry pressing dies. After pressing a cylindrical pellet in a dry press die, it has to be forced out of the die for this very reason.

Removal of all pressure behind a blockage by shutting off the suspension pump will not change the structure of the blockage except to relieve any elastic deformation caused by the high pressures.

The structures of dilatant blockages are very similar to the structures formed in compacts during dry pressing and some filter pressing operations. Dry pressing and filter pressing operations are designed to form strong compacts in desired shapes. Dilatant blockages similarly are strong (but unwanted) compacts that form in the shape of the continuous flow channels.

Detecting Dilatant Blockages in Viscometers

Just as it is difficult to detect dilatant blockages in flow channels, it is equally difficult to detect them in viscometers. It is doubtful such a blockage would ever occur in an infinite-sea type rotational viscometer (see Chapter 14) because the distance between the bob and the container is relatively large and the rotational speeds are limited to relatively low values. Sufficiently high shear rates required to achieve extremely dilatant behaviors are not usually possible in this type of viscometer.

Blockages can easily occur in cup-and-bob and cone-and-plate viscometers (again, see Chapter 14). In these viscometers, blockages occur within the narrow gaps between the rotors and the stators where it is difficult for the operator to see what is happening. When the sensor surfaces are smooth, as they are in many cases, blockages can occur even as viscometers continue their measurements.

Some cups and bobs are profiled with deep groves parallel to the axis of rotation. When these are used, it is possible that the bobs can lock up and rotation can cease. If this happens, the quality of the slip clutch will determine whether the viscometer has been ruined or not. As Professor Funk always joked[8], if the viscometer motor begins to smoke, you know something bad is happening.

Figure 12.1 shows a shear stress versus shear rate rheogram for a dilatant suspension. The problem that can occur in such measurements is due to the fact that viscometers usually show the shear rate created by the rotating cone or bob, relative to the stationary plate or cup, respectively. There is no easy way to measure or to show the actual, instantaneous shear conditions within the suspension. It is simply assumed that suspensions are being sheared at the rates defined by the viscometer.

Figure 12.2 shows the same shear stress versus shear rate rheogram, but in this case, the author has attempted to show the actual shear stress versus shear rate conditions occurring within the suspension when the dilatant blockage occurs.

At the instant a dilatant blockage forms, shear within the suspension stops. If the blockage slides against the rotating bob (or cone) in the viscometer, and the viscometer is actually plotting the

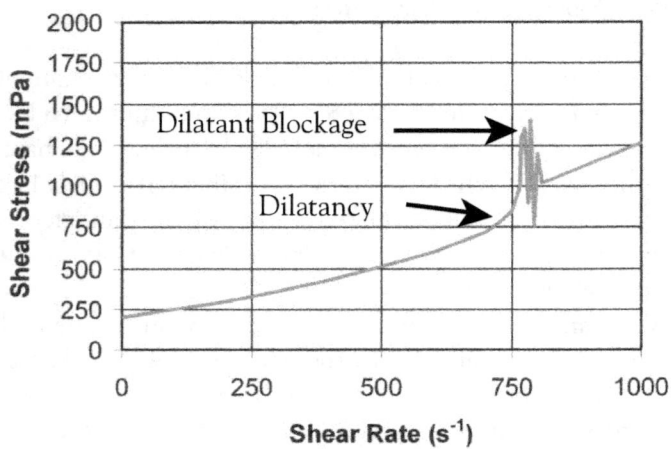

Figure 12.1 A Dilatant Rheogram Showing
Viscometer Shear Rate vs Measured Stress

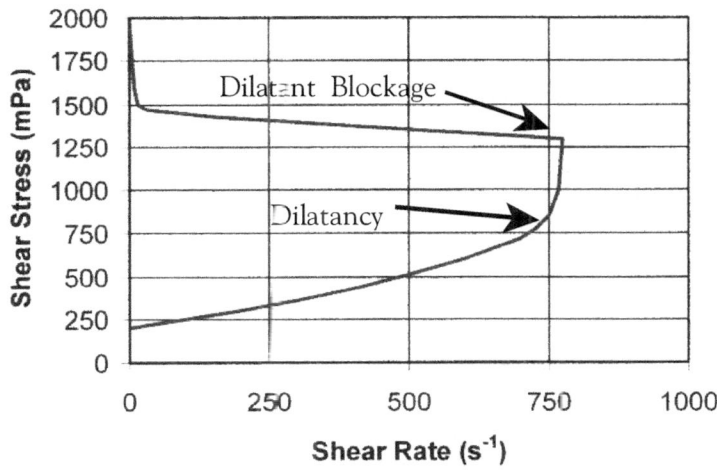

Figure 12.2 A Dilatant Rheogram Showing
Actual Shear Rate within the Suspension vs Measured Stress

shear rate corresponding to the rotating bob's (or cone's) rpm, the viscometer will not record the blockage. It cannot.

If friction between the surface of the blockage and the viscometer, and shear stress from the shearing of the carrier fluid at the surface of the blockage continue to increase as viscometer rpm and imposed shear rates continue to increase, the rheogram could continue as shown in Figure 12.1 at the higher shear rates after the blockage. The nature of the continuation of the rheogram could be Bingham as drawn, or it could continue to chatter around, or it could take some other shape entirely.

If there was a way to look into the suspension to monitor the actual shear that was occurring, a dilatant blockage should produce a rheogram that more accurately resembles Figure 12.2. At the instant the blockage forms, shear stops and returns quickly towards zero. Imposed stresses may continue to build, but shear definitely stops.

It is highly doubtful that a rheogram like the one in Figure 12.2 will ever be produced by a viscometer because to do so, the viscometer would have to detect the actual shear conditions within the suspension.

Theoretically, this is what viscometers are designed to do. But actually, rotational viscometers are designed to produce well-defined shear conditions between a rotor and a stator and **it is assumed** that suspensions or fluids in the measuring are actually shearing at rates consistent with the imposed shear conditions.

In simple fluids, this is not a problem. It only becomes a problem when dealing with suspensions. And it is only a problem in suspensions when imposed shear rates and/or suspension solids contents are high.

Signs That Indicate Dilatant Blockages

When dilatant blockages occur, the surface of the suspension at the edge of the shearing zone in a viscometer changes from a shiny fluid appearance to a dull, matte, dry finish. If the surface of the suspension at the outer edge of the measuring cell is visible to the operator, this change in appearance would be indicative of a dilatant blockage. If the viscometer measuring cell is designed in such a way that excess suspension hides the edge of the measuring gap, or if the edge of the shearing zone is simply not visible to the operator, it may not be possible to see this phenomenon.

If the cup and bob are profiled (large grooves that run from top to bottom on both the cup and the bob) rather than smooth, the viscometer might actually detect dilatant blockages. If so, the viscometer's slip clutch may be called into service, or the stress sensor could be wound up and ruined, or the motor could come to a dead stop, or the motor might start to smoke. Expensive viscometers should be protected by well-designed slip clutches so a lockup should not be a major problem. Read the viscometer documentation carefully (or phone the company and ask) to determine whether this might be a problem. Don't wait until measuring an extremely dilatant sample to find out how the viscometer will handle it (or whether the viscometer can handle it.)

Summary

As dilatant particle/particle interactions increase at high shear rates and high solids contents, dilatant structures can form. When such structures grow to completely fill the cross-sectional area of flow channels, flow will cease. The continued application of high pressures on dilatant blockages causes the blockages to densify and strengthen.

The typical solution to such blockages is to remove and replace the section of flow channel containing the blockage. Simply replacing a blocked pipe with a new, clean section of the same size, however, will allow the process to repeat and a new blockage to form. When such blockages occur, they signify that local shear rates are too high for the process suspension.

Replacing blocked pipes with new pipes of increased diameter will help prevent new blockages from occurring. Alternatively, replacing blocked pipes with new pipes of the same size **and then** decreasing suspension flow rates in that section of pipe can also help to prevent new blockages.

Dilatant blockages can also occur in extrusion and injection molding dies. When flow channel sizes are fixed, local shear rates **must** be decreased to prevent new blockages from forming.

Unfortunately, dilatant blockages, once formed, are not easily dissipated. The general solution to this problem is to operate processes under low shear conditions that do not favor the formation of blockages.

Regarding the detection of dilatant blockages by viscometers, the main points to be emphasized are:

(1) Most viscometers cannot detect the occurrence of dilatant blockages.

(2) Most viscometers continue to record rheograms as if nothing unusual has happened even after a dilatant blockage has occurred.

(3) When conditions are right for dilatant blockages to occur, the viscometer operator has to pay close attention to the suspension surface (during the measurement process) to determine if and when a blockage has formed.

Chapter Thirteen

Practical Rheology

Having gone through the official definitions and explanations of time-independent and time dependent rheologies, a discussion needs to be added to point out some practical issues.

There Is Only One Important Time-Independent Rheology

For ceramists, the only time-independent rheology of real importance to particle/fluid suspensions is the *yield-dilatant rheology*.

Yield stresses are necessary to form ceramic wares because without them, the wares cannot hold their shapes. The yield stress requirement narrows the possible important types of time-independent rheologies for ceramic processing to the three yield-rheologies.

Suspensions that exhibit gel structures and yield stresses are usually shear-thinning (time-independent) and thixotropic (time-dependent). Both rheologies are produced when shear causes breakdown of the gel structure that produces the yield stress.

But since dilatancy is caused by particle/particle collisions, and all ceramic forming bodies and suspensions contain lots of particles, all ceramic bodies and suspensions can (will) also be dilatant at high shear rates. The only way to eliminate dilatancy is to eliminate all possibility for particle/particle collisions. To do this, processes must either be operated (1) at extremely low solids contents, or (2) at extremely low shear rates.

Ceramic suspensions are not practical at the extremely low solids contents required to eliminate all particle/particle collisions. Similarly, ceramic processing is not practical at the extremely low shear rates required to eliminate all particle/particle collisions.

Since neither of these solutions is possible, ceramists must routinely deal with suspensions and forming bodies in which particle/particle collisions (and the potential for dilatancy) occur.

All ceramic suspensions and forming bodies contain particles; particles will interact and collide at sufficiently high shear rates; and particle/particle collisions cause dilatancy. Therefore, it is safe to assume that **all** ceramic suspensions will exhibit dilatant rheologies at high shear rates.

The one rheological requirement and the one rheological consequence for ceramic bodies and suspensions are that they exhibit (1) yield stresses and (2) dilatancy. Only one form of yield-rheology accommodates both of these phenomena: yield-dilatancy.

The one time-independent rheology of importance to ceramists is *yield-dilatancy*.

All Three Yield-Rheologies Are Yield-Dilatant

How can this be? If all ceramic suspensions and bodies are yield-dilatant, how can they also be yield shear-thinning or yield-Newtonian (which is Bingham)? The answer is that most viscometers and rheometers cannot measure apparent viscosities at the high shear rates required to see and record the dilatancy.

The onset of dilatancy in many suspensions is well beyond the upper measurement limit of most common viscometers. (At least that is where we'd like the onset of dilatancy to always be.) The fact that the viscometers cannot measure apparent viscosities at such high shear rates doesn't mean dilatancy doesn't occur. It just means that it doesn't occur within the commonly measured ranges of shear rate.

Many processes subject forming bodies and suspensions to higher shear conditions than can be measured in common viscometers. Because of this, the possibilities for dilatant interactions to occur in almost every case must be considered by process engineers.

Why is this important? The characterization of suspensions as yield shear-thinning and Bingham specifically suggests, by omission, that *these suspensions are not dilatant*. When we, as ceramists and ceramic engineers, do **not** hear the word *dilatant*, we breathe a collective sigh of relief because we 'know' our suspensions will not cause us any problems. When we hear the word *dilatant*, our blood pressures rise and we worry how we can successfully use dilatant suspensions in our processes.

The author has encountered those who aren't moved at all by the thought or threat of dilatancy. Two attitudes that have been expressed to the author are: (1) Dilatancy is not an important nor disastrous rheology to deal with in ceramic process systems; and (2) Dilatancy doesn't exist at all ... rheograms that show increasing apparent viscosity with increasing shear rate are simply the result of poor mixing.

Neither of these ideas is correct, nor should they be bragged about. Ceramic engineers and ceramists **should** get hyper at the suggestion that their process suspensions are dilatant.

The author has also encountered those who ignore particle/particle interactions in favor of mathematical simplicity. The author once observed a technical presentation in which rheograms were shown that were measured over the shear rate range from about $1s^{-1}$ to $100s^{-1}$. The data was limited to very low shear rates due to the particular viscometer used. Someone asked what the apparent viscosity of the suspension would be at the high shear rates in a spray dryer atomizer. Since the suspension was supposed to be Bingham, they were told to simply use a linear extrapolation of the measured rheogram up to the shear rate of the atomizer.

This answer lacked any consideration of the effects particle/particle collisions have at higher shear rates. It incorrectly suggested that particle collisions are not important and can be ignored.

Extrapolation of any rheogram is inherently dangerous. Interpolation, within the range of measured data, can be used successfully, and it can be reasonably informative. But extrapolation using data measured at low shear rates (from $\sim 1s^{-1}$ to $100s^{-1}$) to calculate apparent viscosities at high shear rates (from $\sim 5,000s^{-1}$ to $10,000s^{-1}$ or higher) produces nothing more than a blind guess.

The conclusions to be drawn from these observations are: (1) Some technical people do not consider dilatancy to be a major rheological problem. (2) Some (many??) do not associate particle/particle collisions with dilatancy. (3) Some believe it is not necessary to try to measure viscosities at high shear rates because they can be easily calculated instead.

Those who hold such attitudes are overlooking the real difficulties presented by particle/particle interactions and collisions in particulate suspensions.

If you are a ceramic engineer, technician, artist, or manager, or you hold allegiance to any other field of engineering, and you get hyper at the suggestion that your suspensions may be dilatant ... that's good! Dilatancy is a major problem in suspensions and forming bodies and **it should be taken seriously!**

Shear Rates High Enough?

Figure 13.1 shows three rheograms that all started with shapes identical to that of Rheogram A. Rheograms B and C have

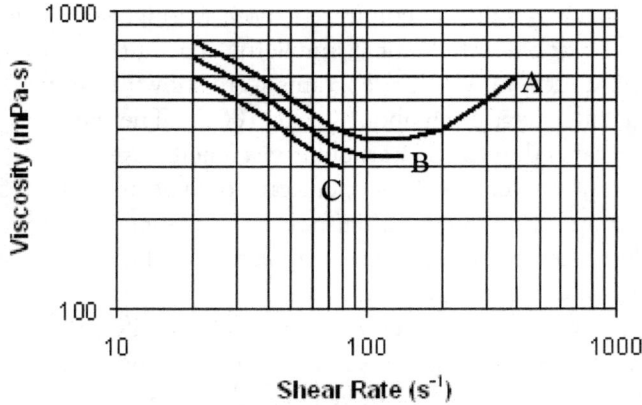

Figure 13.1 Three Yield-Dilatant Rheograms

been mathematically shifted to lower viscosities so they are easier to see.   All three rheograms begin at shear rate $20s^{-1}$. Rheogram A is plotted to $400s^{-1}$. Rheogram B is plotted to $140s^{-1}$ and Rheogram C stops at $80s^{-1}$. Note that Rheogram A is clearly a yield-dilatant rheogram, but Rheogram B appears to be Bingham, and Rheogram C appears to be yield shear-thinning.

All three rheograms, however, are exactly the same yield-dilatant shape (that is, they are all parallel.) Rheograms B and C were mathematically shifted to lower viscosities, and some of the data points in each were not plotted. This figure shows that yield-dilatant rheograms can appear to be both yield-shear-thinning and Bingham, depending upon the upper limit of the measured shear rate.

Many viscometers are limited to relatively low shear rates and they simply cannot be used to measure high shear rate behaviors. As a result, many measured rheograms exhibit shapes that resemble Rheograms B and C in Figure 13.1, even though they would appear to be yield-dilatant rheograms if the measurements could be continued to higher shear rates.

Two questions should be asked when one measures a shear-thinning rheology, or a Bingham rheology: What is the shear rate at this suspension's onset of the dilatancy? Does the measured shear rate range represent all shear rates that occur in the process environment? The answers to these two questions should help to determine whether dilatancy will cause problems for the particular suspension in its process environment.

When suspensions are adjusted by changing chemical additive concentrations, or by making changes to particle size distributions, or by making changes to solids content, one should also wonder how each of these adjustments affects the shear rate at the onset of dilatancy.

If process shear rates are similar to, or less than, the shear rate at the suspension's onset of dilatancy, process adjustments and accommodations should definitely (must??) be made.

Simultaneous Gelation and Particle Interactions

Gelation phenomena and particle/particle interactions occur simultaneously in ceramic process suspensions. Measured apparent viscosities are an equilibrium condition controlled by the gelation rate (which builds the gel) and the shear rate (which breaks the gel and controls the magnitude of particle/particle conditions.)

Gel buildup, gel breakdown, and particle/particle collisions all occur simultaneously in suspensions subjected to shear. The only phenomenon that occurs in quiescent suspensions is gel buildup. Gel is broken down in suspensions subjected to shear by the shear acting on the fluid and gel structure, and by the transfer of energy as remnants of the gel structure, flocs, and particles interact and collide in response to the imposed shear rate.

In low solids content suspensions where collisions and other particle interactions can be ignored, rheological properties due to breakdown of the gel structure will be shear-thinning only. Dilatant effects are not the result of gel breakdown.

Figure 13.2A shows an example of the behavior of an idealized shear-thinning suspension. The higher the imposed shear rate, the lower is the measured apparent viscosity. As shear rate increases, gel structure is

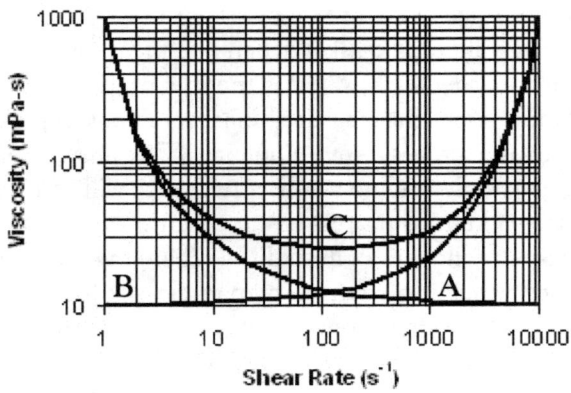

Figure 13.2 Shear-Thinning, Dilatancy, and
Cumulative Effect of the Two

destroyed and apparent viscosity decreases. As shear rate decreases, gelation phenomena dominate, gel structure rebuilds, and apparent viscosities increase. At high shear rates after the gel structure has mostly been destroyed and all particles are traveling as individuals, apparent viscosities are unaffected and relatively constant.

Figure 13.2B shows an example of the behavior of an idealized dilatant suspension. As shear rate increases, particle/particle interactions increase and measured apparent viscosities increase. As shear rates decrease, particle/particle interactions decrease and measure apparent viscosities decrease. At very low shear rates where particle/particle interactions are negligible, apparent viscosities are unaffected and relatively constant.

All of these phenomena occur simultaneously. The third rheogram (C) in Figure 13.2 shows an example of the cumulative effect that results when both types of behaviors occur simultaneously. The cumulative effects shown in this figure are typical of most ceramic suspensions and forming bodies.

At high shear rates where the gel structure has mostly been destroyed, apparent viscosities are high due to the dominant particle/particle collisions and dilatant properties. At very low shear rates where particle/particle interactions are negligible, apparent viscosities are also high due to the dominant gelation phenomena and the gel structure.

Dilatant particle/particle collisions and interactions are disruptive of gel structures, so as solids contents increase, one can expect shear-thinning behavior to end at relatively low shear rates and dilatant properties to dominate over most of the shear rate spectrum.

The point of this figure and this section is to emphasize that all of these phenomena occur simultaneously. The attractive forces between particles cause gel structures to form, and the collisions and interactions between flowing particles cause dilatant properties.

When particles are present in high concentrations, as they are in ceramic process suspensions, gel structures, shear-thinning rheologies, and dilatant rheologies are all present and they all occur simultaneously. The cumulative effects produce yield-dilatant rheologies.

Power Law Equations for Rheograms

One of the common forms of equations used to mathematically represent time-independent rheograms is the power law equation. The general power law form can be used to describe all six time independent rheologies:

$$\tau_s = \tau_y + K \dot{\gamma}^n \qquad (13\text{-}1)$$

where τ_s = shear stress,
K = empirical constant, coefficient of rigidity,
$\dot{\gamma}$ = shear rate,
n = empirical constant, known as the *flow index*, and
τ_y = yield stress.

Equations with flow indices, n, greater than one ($n > 1.0$) characterize dilatant suspensions. Flow indices less than one ($n < 1.0$) characterize shear-thinning suspensions. Flow indices exactly equal to one ($n = 1.0$) characterize the linear behaviors of Newtonian and Bingham suspensions.

Yield stress values, τ_y, equal to zero ($\tau_y = 0.0$) characterize the three non-yield-stress, time-independent rheologies. Yield stress values greater than zero ($\tau_y > 0.0$) characterize the three yield-rheologies. The equations for the six time-independent rheologies are:

Yield-dilatant	$\tau_s = \tau_y + K\,\dot{\gamma}^n$	$(n > 1)$	(13-2)
Bingham	$\tau_s = \tau_y + \mu_B\,\dot{\gamma}$	$(n = 1)$	(13-3)
Yield-pseudoplastic	$\tau_s = \tau_y + K\,\dot{\gamma}^n$	$(n < 1)$	(13-4)
Dilatant	$\tau_s = K\,\dot{\gamma}^n$	$(n > 1)$	(13-5)
Newtonian	$\tau_s = \mu\,\dot{\gamma}$	$(n = 1)$	(13-6)
Pseudoplastic	$\tau_s = K\,\dot{\gamma}^n$	$(n < 1)$	(13-7)

Note that the proportionality constant, K, used for Newtonian fluids is known as the Newtonian viscosity, μ, and the one used for Bingham fluids is known as the Bingham viscosity, μ_B.

    These six rheograms are shown in a shear stress versus shear rate diagram in Figure 4.1 and in an apparent viscosity versus shear rate diagram in Figure 4.2.

    The power law equations of the non-yield stress rheologies plot as straight lines on log-log axes as shown in Figure 4.2. The power law forms of the yield-rheologies approach linear behavior at higher shear rates. Because of this linear behavior, log-log axes are commonly used for rheograms.

    Log axes are useful for both axes because shear rate, shear stress, and apparent viscosity can all cover several orders of magnitude. Log axes allow easy plotting and reading of shear rates, shear stresses, and apparent viscosities over the wide range of shear rates represented in a single graph.

    For example, in a log-log plot, apparent viscosity behavior in the 1s^{-1} to 10s^{-1} shear rate range is as easy to see as the apparent viscosity behavior in the $1{,}000\text{s}^{-1}$ to $10{,}000\text{s}^{-1}$ range. A normal (linear) axis covering the shear rate range from 0s^{-1} to $10{,}000\text{s}^{-1}$ would be dominated by the data between $1{,}000\text{s}^{-1}$ and $10{,}000\text{s}^{-1}$. The low shear rate data, which would be compressed into the region near the viscosity axis, would be difficult to read.

Figures 13.3 and 13.4 show examples of this. Figure 13.3 shows the shear-thinning rheogram of Figure 13.2A plotted on log-log axes. Figure 13.4 shows that same rheogram plotted on semi-log axes.

In Figure 13.3, it is quite easy to read apparent viscosities in the $1s^{-1}$ to $10s^{-1}$ shear rate range. In Figure 13.4, however, it is nearly impossible to accurately read any apparent viscosities below about $500s^{-1}$. For this reason, log axes are recommended.

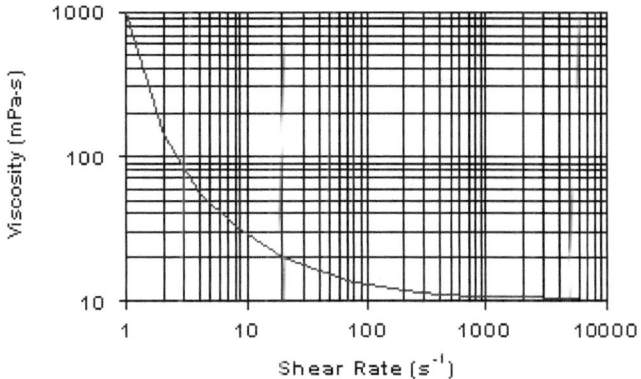

Figure 13.3 Shear-Thinning Rheogram
Plotted on Log-Log Axes

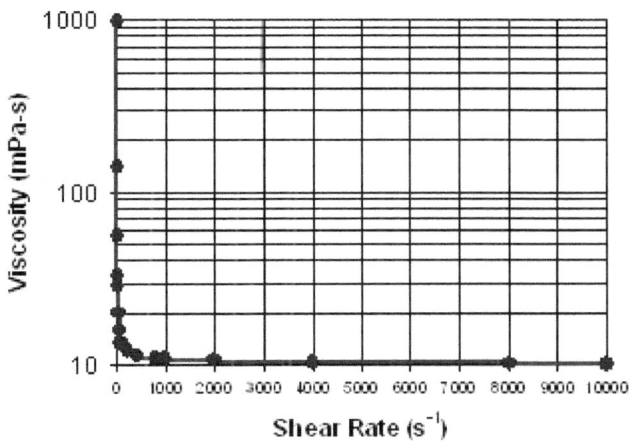

Figure 13.4 Shear-Thinning Rheogram
Plotted on Semi-Log Axes

What to Expect at High Shear Rates?

Figure 13.5 shows a log-log presentation of a perfect shear-thinning rheogram of the form of Equation (13-7). It has a perfect linear behavior on log-log axes.

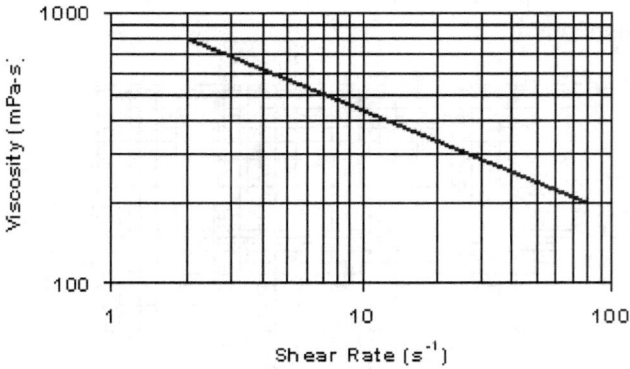

Figure 13.5 A Perfect Pseudoplastic Suspension

The questions that need to be asked are this:

1 – What is the viscosity of the carrier fluid in this suspension?
2 – How far can this linear behavior continue? That is, to how low a viscosity can this linear behavior continue as shear rates continue to rise?
3 – Since log-log axes have no zero values, will the viscosity continue to decrease towards, but never reach zero, as shear rates rise?
4 – Will the suspension viscosity eventually drop below that of the carrier liquid alone?

The answer to the first question sets a reasonable limit to the discussion of viscosity. Let's assume the viscosity of the carrier fluid alone is about 100 mPa·s.

How far will the linear behavior continue as shear rates rise? If no easy answer is forthcoming, consider questions 3 and 4. As shear rates rise, will the viscosity of the suspension eventually approach zero? Will the viscosity of a powder/water suspension eventually decrease below the viscosity of water? If the viscosity of the carrier fluid in Figure 13.5 is 100 mPa · s, will the viscosity of the suspension eventually decrease below 100 mPa · s? Will the viscosity of a powder/water suspension, or the viscosity of the suspension in Figure 13.5, eventually decrease below the viscosity of air? Expressed in this way, the suggestion that the linear viscosity behavior will continue ad infinitum (which it can do mathematically) is absurd.

    The suggestion that the viscosity of a suspension could eventually decrease below the viscosity of the carrier fluid alone, is absurd. No! It can't. Shear alone can't reduce the viscosity of water below the viscosity of water. Neither can shear alone reduce the viscosity of a powder/fluid suspension below the viscosity of the carrier fluid.

    Einstein's equation (Equation 3.1) shows that the viscosity of a low solids content suspension is equal to the viscosity of the fluid **plus** an effect (one that increases viscosity) for the solids in suspension. Lots of empirical data are available to demonstrate the veracity of Einstein's equation to low solids content suspensions.

    So the rheogram shown in Figure 13.5 must eventually turn from its linear behavior to approach a limiting viscosity that can be no lower than the viscosity of the fluid alone. But that produces a rheogram that has Bingham character, as shown in Figure 13.6.

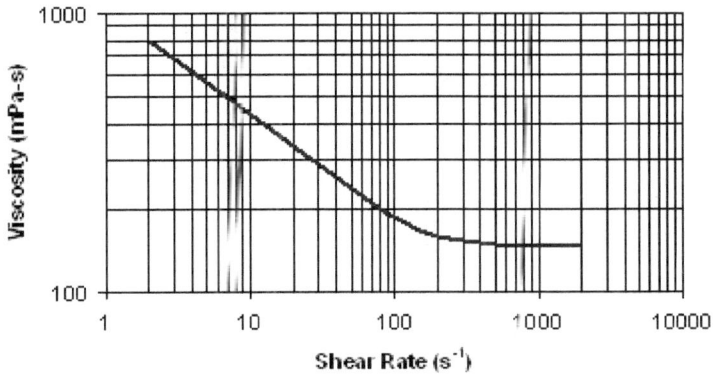

Figure 13.6 A Bingham Rheogram? The rheogram in Figure 13.5 reaches a limiting apparent viscosity.

Is this rheogram Bingham? To fit the Bingham behavior (Equation 13-3), the viscosity must level off at some point as shear rates continue to rise and approach constant viscosity.

To continue this series of questions, we must now ask:

5 – Will particles ever begin to collide as shear rates
continue to rise?

If the answer to Question 5 is 'No':

6 – Why not?

If the answer to Question 5 is 'Yes':

7 – Will the intensity of collisions rise as shear rates
continue to rise?
8 – Will collisions eventually dominate as shear rates
continue to rise?

These questions ask whether the rheogram will continue as Curve A, or as Curve B, in Figure 13.7.

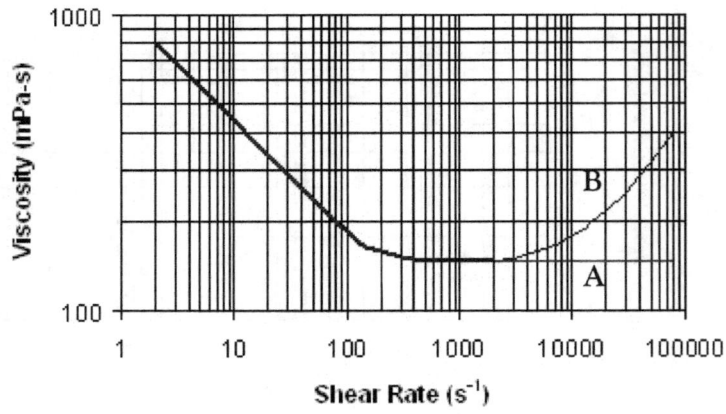

Figure 13.7 Continuation of Rheogram as Shear Rates Rise

If the suspension is truly Bingham, and if collisions never intensify or dominate within the suspension, then a continuation of the rheogram should follow Curve A. But if collisions will intensify as shear rates increase, and if, eventually, collisions will dominate during high shear conditions, then the rheogram will follow Curve B. Curve B, however, is no longer a Bingham rheology, but it is characteristic of the yield-dilatant rheology.

If this discussion concerned extremely low solids contents, such as 1% solids and 99% fluid, one might make the case that collisions between the particles would never dominate the rheology. This might be true.

But typical ceramic suspensions are fairly concentrated, sometimes 50 volume% solids or more, and typical ceramic forming and extrusion bodies are even higher. Particles in such suspensions and bodies will collide and dilatancy will not only appear, but it will dominate at 'high' shear rates.

Curve B in Figure 13.7 is the type of rheogram one should expect for ceramic process suspensions and forming bodies. The only question that remains for each ceramic process environment is: What is a 'high' shear rate in this process? Each ceramist has to answer this question for his or her own process.

Each production process will have some high shear operations and many low shear operations. The low shear operations are not the problem. Identifying the high shear operations, and characterizing the imposed shear rates in those operations, is the problem.

All who are responsible for the successful operation of a production process must be familiar enough with the process to know the locations of the high shear operations in that process, as well as the process slips' and bodies' expected behaviors in those locations.

Whether one person, or several, are responsible, this type of information on the processes, suspensions, and bodies in each ceramic plant should be known and closely monitored.

To ignore particle/particle interactions, or to suggest that dilatancy doesn't exist, is risky. Dilatancy has caused many, many processing problems, and its effects can be extreme.

Practical Suggestions Concerning
 ## Apparent Viscosity Measurements

High Shear Measurements

It is difficult to measure high shear behaviors in many ceramic suspensions. Many good, industrial viscometers simply can't reach sufficiently high shear rates to show dilatant behaviors. Many of the viscometers that can reach high shear rates are expensive and are not readily available for process control.

It is equally difficult to measure high shear behaviors of ceramic forming bodies. Not only are the rheometers, that are required to make such measurements, relatively scarce within ceramic production companies, but these types of rheometers are also frequently limited to relatively low shear rates.

Measuring Rheology vs Apparent Viscosity

To learn the complete rheological behavior of a process suspension or forming body, one must measure apparent viscosities at **more than one** shear rate. If a rheometer is always run at a single rpm, the measurements will indicate apparent viscosities at this rpm, but the measurements will not indicate anything about the rheological types of the samples.

To measure **rheologies**, (as opposed to measuring apparent viscosities) suspensions should be tested at several shear rates (that is, at several rpm settings). **At least two** different rpms must be used to learn anything about the rheology of a suspension.

Rpm is approximately proportional to shear rate, so even if the shear rate at each rpm is not precisely known, tests performed at 50, 100, and 200 rpm will yield rheological information. Each test rpm reveals apparent viscosities at that rpm, but two or more rpms reveal rheological information.

When apparent viscosities are measured at two or more rpms (which corresponds to two or more shear rates), rheograms can be plotted as shear stress versus rpm. If the range of test rpms covers the viscometer's full range, the results will be the best possible from that particular viscometer. One of the shear rates tested in each case should include the maximum possible shear rate (rpm) measurable by the viscometer.

Dynamic vs Kinematic Viscosity Measurements

Note that the measurements we need to make are dynamic viscosity measurements. Dynamic viscosities can be measured by cup-and-bob, rotating cylinders, cone-and-plate, vibrating spheres, and capillary viscometers.

Kinematic viscometers are those in which suspension flows are timed through an orifice in the bottom of a cup, or through an orifice in tubing, under the influence of gravity alone. Kinematic viscometers do not yield useful rheological information. Such viscometers may be useful to tune and adjust process viscosities from batch to batch, but they are not useful to determine rheologies.

The types of viscometers that can measure dynamic viscosities over a wide range of shear rates are characterized by the capability to expose all of the suspension in the measuring cell to a single shear rate. To measure rheological properties, several measurements from several different shear rates are used.

Because kinematic viscometers use gravity to produce the flow and shear conditions, each measurement usually represents the cumulative effects of a range of shear rates. The pressure head, created by the height of fluid from the orifice to the fluid surface, controls the flow velocity (and the shear rate) of fluid passing through the orifice. As fluid passes through the orifice and the surface level drops, the pressure head, the flow velocity through the orifice, and the shear rate within the orifice all decrease. For these reasons, a single kinematic viscosity measurement represents a range of shear conditions.

This is not to say that kinematic viscometers are not accurate, reproducible, or useful. They are all of these. They just don't produce useful information about the **rheological** properties of a suspension.

Automatic Viscometer Measurements

Many viscometers today are controlled by computers. It is relatively easy to set up such a viscometer to measure apparent viscosities at several different shear rates from the lowest to the highest possible.

Automatic viscometers are excellent from the point of view of consistency. Since most ceramic suspensions are thixotropic (time-dependent), it is important that apparent viscosities of individual samples be measured using identical procedures. Computer-controlled viscometers are excellent at reproducing identical test conditions from sample to sample.

Manual Rheology Measurements

If a computer-controlled instrument is not available, a stop watch should be used to precisely control the duration of measurement at each different shear rate. When manually performing rheological measurements, rpms should be changed as the viscometer continues to run and measure data.

A 10-minute rheology run that starts at a low rpm and measures 10 apparent viscosities at 10 different rpms should take exactly 10 minutes. At the end of each 60 seconds, the rpm should be increased to the next value. Unless forced to do so by viscometer requirements, don't stop the viscometer at the end of each 60 seconds to change the rpm and then restart it to continue the test. Simply take the reading at the end of each 60 seconds and then immediately increase the rpm to the next value with the viscometer running.

A Test to Measure Rheology

An acceptable program for measuring the apparent viscosities at 1, 5, 10, 50, and 100 rpm, is to allow the viscometer to measure viscosities at each rpm for precisely 1 minute. The apparent viscosity at the end of 1 minute at each setting should be recorded, and the rpm should be increased precisely at the end of each minute. The overall test will then take 5 minutes to perform.

If the viscometer can measure any rpm between 1 and 500, a good test would be to measure apparent viscosities at 10, or 20 different rpms after durations of 30 or 60 seconds at each rpm. Pick a single time duration at each rpm and then use it from test to test, day after day. A 60 second duration at each rpm is good, but each test to measure apparent viscosities at 20 different rpms will take 20 minutes. If that is too long, then select a 30 second duration at each rpm instead.

When a broad range of shear rates is possible, the author recommends selecting the rpm values that fit a log scale. If the rpm range is 1 through 500, instead of selecting 1, 50, 100, 150, and so on by 50s to 500, select rpms that produce uniform increments on a log scale. Table 13.1 shows how to calculate such a scale to divide the range from 1 to 500 rpm into 10 even log intervals.

The logs of the lowest and highest rpms (1 and 500) are placed in Column 2. The difference between these two log values, 2.70 - 0.0 = 2.70, is divided by 9 to determine the incremental value (0.30). Note that the number of intervals is one less than the number of rpm values. For 10 rpm values, use

9 intervals. Starting with the first row, add this increment to calculate each new log value as shown in Column 3. Take the anti-logs of these values, shown in Column 4, and round these off to produce the new rpm values shown in the last column. When plotted on a log rpm axis, the values in the 5th column of Table 13.1 will produce evenly spaced increments.

   To complete the test setup, the decision must be made concerning the duration of the measurement at each test rpm, which corresponds to the number of seconds between measurements. Since this example has 10 rpm values, the test will take exactly 5 minutes when 30sec durations are used, and 10 minutes when 60sec durations are used. Program this information into the viscometer, and every time a rheogram is needed, run this procedure.

   It is not necessary to use log increments. It is not necessary to use evenly spaced linear increments. It is not necessary to use 30sec or 60sec

Table 13.1 Calculating a log scale between 1 and 500 rpm.

rpm	log rpm	log rpm interval	anti- logs	new rpms
1	0.00	0.00	1.0	1
		0.30	2.0	2
		0.60	4.0	4
		0.90	7.9	8
		1.20	15.8	16
		1.50	31.6	32
		1.80	63.1	63
		2.10	126.	126
		2.40	251.	251
500	2.70	2.70	500.0	500

time intervals. Either is okay. Log increments are recommended because they are handy and provide information uniformly across the shear rate spectrum.

   One thing is necessary, whatever the decision. The same procedure must be used **consistently** from test to test. After a rheological measurement procedure has been designed, whether it's programmed into a computer-controlled viscometer or it's going to be run manually, it needs to be followed exactly and precisely each time. Only then will measurements correspond to one another and only then can differences between samples be compared.

Summary

The only time-independent rheology of importance to ceramists is the yield-dilatant rheology. If a measured rheogram appears to be shear-thinning, or Bingham, one should wonder where the onset of dilatancy occurs.

Gelation, which builds gel structures, works to reform gel structures even as they are being broken down by shear. Particle/particle interactions and collisions cause dilatancy as they help to disrupt and breakdown gel structures. All of these phenomena occur simultaneously in process suspensions. Measured apparent viscosities exhibit the cumulative effects of all of these phenomena. When a state of constant apparent viscosity has been achieved at a particular shear rate, all of these competing effects have reached a state of equilibrium.

All ceramic process suspensions should be considered to be yield-dilatant. The main question of interest for each suspension should concern the shear rate at which the dilatancy begins to appear. Trying to decide whether or not a suspension is dilatant is a waste of time. Just assume that all suspensions are yield-dilatant and try to determine the shear rate at which the dilatancy will appear.

In many circumstances, dilatancy will begin at shear rates well beyond the highest represented in the process environment. In some cases, however, process shear rates will be greater than the shear rate at the onset of dilatancy. Process problems will then occur.

Log-log axes are recommended when charting rheograms. Apparent viscosity data over the wide range of available shear rates are well-displayed and easy to read when log axes are used.

To measure rheological properties, viscometers that can measure dynamic viscosity must be run at several different rpm values. Measurements at only a single rpm are simply **apparent viscosities** at a single shear rate and they say nothing concerning a suspension's rheological type. Viscometer measurements covering several rpms (i.e., at several different shear conditions), however, will provide data that characterizes suspension **rheology**.

Chapter Fourteen

Viscometers and Rheometers

Several types of viscometers are available to measure rheological properties. This chapter will cover some of the different types of viscometers that can measure dynamic viscosities and rheological properties.

Each type of viscometer has specific considerations that should be applied as tests are run and as data is interpreted. These considerations will be mentioned and discussed.

Rotational Viscometers

The most common type of viscometer used in the ceramics industries to measure viscosities and rheologies is the rotational viscometer. A cylindrical spindle (a bob) is inserted into suspensions to measure viscosity. Rotational viscometers drive the spindle and measure the resistance of the suspension to the spindle rotation. The spindle is usually linked by a spring or torque sensor to the drive motor. This allows viscosity measurements to be made as the spindle revolves.

Different types of rotational viscometers represent different measurement configurations and geometries. The different geometric types, infinite sea, cup-and-bob, and cone-and-plate, will be covered separately in the next sections.

Infinite Sea Viscometers

The name of this type of rotational viscometer comes from the fact that the bob can be inserted into any size container from a small beaker, to the ocean itself. The assumption is made that the measure-

Viscometers and Rheometers

ment container is large enough that edge effects are negligible to the measurement data and therefore can be ignored. The measurement container is assumed to be as large as an 'infinite sea.'

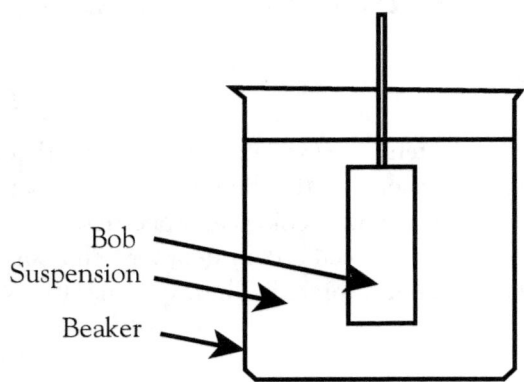

Figure 14.1 Infinite Sea Rotational Viscometer

Figure 14.1 shows a diagram of such a viscometer. If the diameter of the bob is about an inch or smaller, a 400ml beaker works well to hold the suspension. This provides a relatively large gap between the bob and the beaker walls, so edge effects should be minimal. As the spindle rotates, the suspension will be sheared between the outer surface of the spindle and the inner surface of the container. The suspension will see a particular set of shear conditions defined by the bob/container combination used for the measurement. To keep these conditions constant from test to test, a single container size should be used with each bob for all tests.

This type of viscometer is commonly available and used frequently within ceramic companies. These viscometers usually offer a variety of rotational speeds as well as a variety of spindles to cover wide viscosity ranges and measurement types.

RPM as the Unit for Shear Rate

There are equations that purportedly calculate the shear rates for particular spindle/rpm conditions with this type of viscometer. The applied shear rates in all parts of the 'infinite sea,' however, are not constant. As a result, it hardly seems worthwhile to calculate an applied shear rate that is not universally applicable. For this reason, the author simply uses **rpm** as the units for shear rate for this type of viscometer. The rotational rpm is well defined and controlled by the viscometer. Although the shear rate applied to the suspension by the spindle, as well as the calculated shear rate, may not be constant, the rpm value is accurate.

The applied shear rate is generally proportional to rotational velocity. So when the rpm doubles, the shear rate doubles, etc. Therefore, using rpm as the unit for shear rate is a reasonable substitute.

A 100 rpm measurement from this type of viscometer is in the ball park of shear rates that occur within pipe channels. Such measurements are certainly at shear rates well below those imposed within spray dryer atomizers. But this rpm represents a typical shear rate suspensions will see as they are pumped around a process plant.

Lower rpms are indicative of viscosities in slowly agitated holding tanks. The very lowest rpms are indicative of the most quiescent phenomena, such as the viscosity of a body slip as it is casting in a mold.

The key words here are *ball park*. If an apparent viscosity is required at a very precisely controlled shear rate, this type of viscometer is not recommended. But to measure viscosities over typical process shear rate ranges (those in the right ball park), and to measure rheological properties in those same ranges, this type of viscometer is excellent.

Some Pointers

Although edge effects can be ignored with this type of viscometer, different sizes of containers do combine with the rotating spindles to impose different conditions on the suspensions. A measurement taken in the center of the surface of a 55 gallon drum should not be expected to exactly correspond to the measurement of the same fluid or suspension taken in a small beaker. They should, of course, be quite similar, but they may not be identical.

The recommendation is to select a single beaker size and then use it consistently. If suspensions are mixed in a milkshake mixer before each measurement, the milkshake mixer cup can be the standard container in which measurements are made. Or a 400ml beaker can be used. Or some other container may be appropriate. Choose a container and then use that same size and configuration all the time.

For the same reason, different spindle configurations also combine with the container to produce different measurement conditions. Even when the measurement ranges of spindles overlap, two different spindles will not necessarily produce the same viscosity measurement at the same rpm. A particular spindle should be designated as the standard spindle to use for all tests. If different spindles must be used, a data base should be set up for each spindle so measurements can be compared under similar test conditions.

When the spindle configuration has the spindle completely submerged in the suspension and only the shaft breaks the surface, don't forget to put a drop of water at the point where the spindle shaft breaks through the surface. On tests of long duration, surface bubbles, scum, drying suspension, etc., can attach to the shaft and cause readings to increase with time. The goal is to measure the viscosity of the bulk suspension using the surface of the submerged spindle. Increased drag caused by suspension surface structures attached to the shaft can produce erroneous measurements.

Cup-and-Bob Viscometers

A step up from the infinite sea type of viscometer is the cup-and-bob viscometer. In this type of viscometer, the sizes and geometries of both the cup and the bob are precisely controlled.

The cups tend to be much smaller than the containers used in infinite sea viscometers. Shear rates in this type of viscometer can be quite high, and fluids, suspensions, cups, and bobs can all heat up as measurements are taken. Frequently, therefore, constant temperature water jackets are available. Figure 14.2 shows a diagram of a cup-and-bob viscometer.

Constant Shear Rate

These types of viscometers are usually designed so the shear rates imposed on all of the fluid or suspension in the measurement zone (the narrow gap between the cup and the bob) is constant. If a measurement from this type of viscometer indicates that it applies at $100s^{-1}$, all of the suspension in the measurement zone was exposed to $100s^{-1}$ at the time of the measurement.

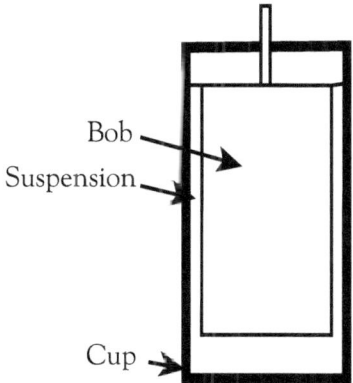

Figure 14.2 A Cup-and-Bob Viscometer

At least that is what the theory says. Each manufacturer should have complete explanations of the viscometers, accessories, and the theory behind their instruments in the documentation distributed. Refer to that documentation for specific details on specific cup/bob combinations on each instrument.

Without going into detail on the theory, it is necessary that a few things be pointed out. The theory says that the shear rate in the gap between the cup and bob approaches constant values as the gap becomes small. These instruments are usually built around this assumption that the gap is very small, so all fluid in the measurement zone is exposed to the same, constant, instantaneous shear rate.

This is fine when the fluid to be measured is a liquid. But what happens when a suspension is to be measured? Well, coarse particles and fine gap sizes don't go together very well. So to prevent such problems, the gap sizes are frequently increased when suspensions are to be measured.

Under such conditions, shear rates are not quite constant. The range of applied shear rates within the measurement cell, however, should be relatively small, so the constant shear rate assumption should still be acceptable. But this small distinction is usually overlooked. Reports and presentations frequently contain rheograms that show apparent viscosities measured at shear rates labeled at a single, precise shear rate value. Some of these measurements actually represent the apparent viscosities measured at shear rates that include a small, but finite, range around the stated value.

This is not to suggest that such results are bad, erroneous, or not useful. This is simply to point out that measurements made on suspensions in cup-and-bob viscometers may not be made under the exact, ideally precise conditions implied by the theory.

Such data are quite useful and frequently, they represent the best measurements possible on suspensions, especially at the high shear rates which these instruments are capable of achieving.

Some Pointers

Because cup-and-bob viscometers can reach really high shear rates, make sure the unit has a good slip-clutch to protect the motor if and when the suspension being measured goes dilatant. The last thing anyone wants to do is to ruin an expensive rheometer with a dilatant suspension.

Whenever possible, watch the surface of the suspension while measurements are made at high shear rates. It is not possible to view the suspension as it is sheared between the cup and the bob, but it is sometimes possible to see the surface of the suspension at the uppermost point of the cup-to-bob gap. If this surface changes abruptly from looking moist, when being sheared, to dry with little detectible movement, it may be an indication that the suspension between the cup and bob has reached the state of a dilatant blockage.

If this occurs, the suspension may not be shearing at all. If a blockage has formed, it may simply be sliding between the cup and the bob. The rheogram from this point on may appear Bingham in nature. If a blockage has occurred between the cup and bob, the force of friction from the sliding blockage will increase as the shear rate increases. The shearing of carrier fluid at the walls will cause the shear stress to increase as shear rate increases as well. Both of these two phenomena may cause the measured stresses to increase linearly with increasing shear rate and therefore the suspension can appear Bingham. But if the particles in the 'suspension' are locked in position and are not being sheared at all, such measurements are meaningless.

It is not possible to see what's actually happening to the suspension in the measurement cell of a cup-and-bob viscometer. If a blockage does occur, but goes undetected or unobserved, data interpretation will be incorrect.

Another point of consideration applies to the measured viscosities as the rotation just begins at the start of a measurement. If the viscometer monitors and plots the rpm of the drive motor rather than the actual rpm of the cup or bob (whichever is rotating), errors will occur at the start of the measurement.

The author conducted an experiment on a cup and bob viscometer which showed this problem. After defining a relatively slow acceleration of the bob from zero to some nominal rpm value, the author held the bob so it could not rotate. The indicated rpm increased steadily due to the programmed acceleration, even though the bob was not moving.

This particular viscometer monitored the drive motor rpm rather than the bob rpm. The assumption was made that the two would always be equal. This is generally true and usually, it's not a problem. This would be a problem, however, if the viscometer was used to monitor and measure gelation behavior or gel strengths. If the suspension to be measured is prepared, poured into the measurement cell, and then allowed to sit quiescent for 20 minutes before the actual run begins, the measured values in the first few seconds of the run (until the gel structure breaks and the suspension begins to see shear) will be incorrect.

This may not be a problem at all in labs where gelation behavior is not of concern. Nor will this problem apply to all viscometers of this type. If the goal for a particular instrument, however, is to measure gel structures, this point needs to be considered.

Cone-and-Plate Viscometers

Cone-and-plate viscometers are based on the simple geometry of an inverted cone with its vertex at the surface of a plate. The real beauties of this type of viscometer are twofold. (1) They require only very small samples; and (2) All of the fluid or suspension in the measuring zone sees one constant shear rate. The small sample size is a real plus. Not much suspension is needed, and cleanup is rather simple. Figure 14.3 shows a diagram of a cone and plate viscometer.

Cone-and-plate viscometers tend to be expensive. This type of viscometer, however, can achieve relatively high shear rates, and the plates can usually be cooled to maintain temperature uniformity throughout the small sample.

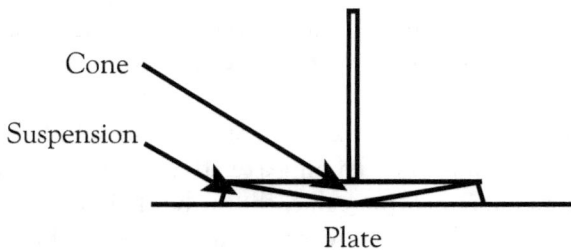

Figure 14.3 A Cone-and-Plate Viscometer

Similar to the theory behind the geometry of cup-and-bob viscometers, the cone-and-plate theory says that when the included angle between the cone and plate is really small, that is, fractions of a degree, and the tip of the cone is very close to the plate, all of the fluid between the cone and plate will see a single shear rate at each rpm. Again, this works well when measuring simple fluids.

The small cone-to-plate angle required is much smaller than that shown in the diagram. When the cone is designed to produce such a small included angle between the cone and the plate, it more closely resembles a flat disc than a cone. The volume of sample contained in the measuring zone is quite small compared to either of the other two rotational viscometers already discussed.

In practice, one difficulty is to maintain a small, and relatively constant, distance between the cone and the plate. Different manufacturers handle this differently. This is necessary, however, to achieve constant shear rate within the total volume of the sample.

But when a suspension is to be measured, if the tip of the cone is close to the plate (as it should be), the possibility exists that particles will bridge the gap and measurement errors will occur. So the solution to this is typically (1) to increase the distance between the tip of the cone and the plate, or (2) to truncate the cone. To truncate the cone is to grind off the tip so it becomes a flat. Both techniques cause suspension in the measurement zone to see more than one shear rate at each rpm. The suspension at the center sees a relatively low shear rate, while suspension near the perimeter of the cone sees higher shear rates.

Again, these are necessary modifications to allow cone and plate viscometers to measure viscosities and rheologies of particulate suspensions. They allow these devices to be applied to a wide variety of fluids and suspensions. Just be aware that when either of these conditions are used to accommodate suspensions in cone-and-plate viscometers, the indicated shear rate values are representative of a small range of shear rates, rather than a single, exact value imposed upon the whole measurement volume.

Dilatant blockages can occur in cone-and-plate viscometers as well as in cup-and-bob viscometers. The same comments and suggestions mentioned for cup-and-bob viscometers apply to cone-and-plate viscometers on this point as well.

Capillary Viscometers

Rather than measuring viscosities on a viscometer and trying to translate the results to process conditions, it would be simpler in many ways to pump the suspension around the actual plant piping and measure the appropriate pressure drops. This has lots of advantages and several drawbacks.

Capillary viscometers are a step in this direction. Fluids and suspensions are pumped into long tubes of very small diameter, and the pressure drops required to achieve the various flow rates are measured. Different flow rates cause different shear rates within the capillary tubes, so rheologies can be measured when two or more flow rates (shear rates) are used.

These viscometers are much less flexible than any of those mentioned above. In particular, they are susceptible to dilatant blockages, and if one forms, the viscometer tube is ruined. If this type of viscometer is used to measure suspension properties, it must necessarily be used at low shear rates.

There are excellent reasons for using capillary viscometers. For example, high shear rates can be achieved in this type of viscometer. But to measure suspensions at high shear rates in capillary viscometers is not advisable. The tubes are just too expensive to allow frequent replacement due to blockages.

Summary

Although other types of viscometers are available, the most common types used for measuring rheologies of suspensions are the rotational types.

Each have their particular advantages and disadvantages. Some are sturdy and can stand up to process environments. Others are more delicate and more expensive.

The author recommends the general category of rotational viscometers for use with ceramic suspensions. The decision and specification of the type, manufacturer, and model that is best suited for any particular lab or production environment remains within the purview of each company.

Chapter Fifteen

Control of Suspension Rheology

A treatise on rheology for the ceramist would not be complete without a chapter on the control of suspension rheology. Controls take two forms: (1) particle physics controls, and (2) chemical controls. For 'optimum' rheological properties, **both** categories need to be controlled, adjusted, and optimized.

The word *optimum* implies that rheological properties are perfect or ideal. This is never the case. Everyone lives with suspensions that do not have fully optimized properties. So don't expect any suspension to ever be completely optimized. When rheological properties must be tightly controlled, however, both categories of adjustments must be taken into account.

These two categories will be discussed individually in this chapter.

Particle Physics Controls

The subject of particle physics includes physical properties of a suspension such as solids content, particle size distribution, surface area, packing capability, etc. The abilities of suppliers to ship powders with constant properties, and the abilities of the engineering staff to accurately weigh and formulate batches are usually relied upon as particle physics 'controls.'

When this author refers to the control of particle physics, he assumes that each of the important properties of all of the raw material ingredients are (1) being measured and (2) being altered and/or adjusted as necessary, prior to batching, to achieve constant production body properties.

Most of the control of particle physics properties in production batches occurs up to the point that the powders are mixed into the batch. If the particle size distribution of a powder or the specific surface area of that powder is incorrect, production engineers have the opportunity before batching to make changes. They can also blend powders from several shipments to achieve the desired properties. This applies to any physical property of any batch ingredient.

After all ingredient powders have been added to the batch, however, particle physics adjustments take a back seat to chemical adjustments. After batching, particle physics adjustments (other than solids content) are seldom employed. When holding tanks are used and tank capacities allow, several different production batches can be blended to achieve target properties.

In the next sections, the effects of several of the important particle physics properties will be discussed. These properties can all be adjusted prior to batching, and even after batching (with more difficulty) when necessary.

Particle Size Distribution

The particle size distribution of the powder in a ceramic body is one of the most important properties for a ceramist to control. Many forming properties, firing properties, and fired properties are the result of proper particle size distribution controls. The relationships between all of these properties and particle size distribution combine to form an enormous subject. Some of these relationships which apply to rheological properties will be discussed briefly here.

Packing

One of the most important process properties affected by the particle size distribution (PSD) of a powder is its capability to pack.

Generally speaking, the broader the PSD, that is, the greater the range of sizes represented in the powder, the better the particles can pack. Very narrow distributions, in which all particles are essentially the same size, pack poorly. Obviously, there are many more important details regarding particle packing than just breadth of distribution. Particle packing is a large, complex subject, which will be (and has been[10]) discussed in more detail elsewhere.

The fundamental capacity of the powder fraction to pack densely affects the viscous and rheological properties of suspensions.

Viscous and Rheological Properties

When all particles in a distribution are essentially the same size (which is described as a *narrow* PSD), packing will be poor. Whenever a PSD cannot pack well, solids contents must remain relatively low to achieve acceptable suspension viscosities. Suspensions formed from powders with particle size distributions that pack well can achieve acceptable viscosities at much higher solids contents.

Narrow PSDs that don't pack well typically produce suspensions with dilatant rheological properties even at low solids contents. When such suspensions are forced to the highest possible solids contents, high levels of deflocculation are required to minimize viscosities, and dilatant properties can be extreme.

PSDs that pack well can produce the whole range of rheologies, from shear-thinning to dilatant, depending on the solids contents and the nature and concentration of chemical additives utilized. High solids content suspensions (considerably higher than can be achieved with distributions that don't pack well) may require deflocculants to achieve target viscosities. High solids content, highly deflocculated suspensions will still tend to be dilatant. Low viscosity, low solids content suspensions may require flocculants to achieve production viscosity targets. Such suspensions can be expected to be shear-thinning.

When particle size distributions pack well, a broad range of solids contents can be used to produce suspensions with a wide range of viscosities and an equally wide range of rheological properties (from extremely shear-thinning to extremely dilatant). When particle size distributions don't pack well, solids contents are usually low, viscosities can vary, but rheologies tend to be dilatant.

Carrier Fluids and Packing Potential

There are two uses for carrier fluids in suspensions: (1) Carrier fluids fill pores; and (2) All available carrier fluid that was not required to fill pores (the non-pore carrier fluid) imparts fluidity. These two functions combine with the packing potential of a powder to control suspension viscosity as a function of solids content.

Take, for instance, a powder distribution that packs to 60 vol % packing factor, which defines a porosity of 40 vol %. A 'suspension' of this powder at 60 vol % solids content would not have any fluidity because the 40 vol % fluid would just exactly fill all the pores. Particles would still touch, and there would be no fluidity and no measurable viscosity.

A powder that can pack to a 90 vol % packing factor, which produces 10 vol % porosity, would be very fluid in a 60 vol % solids suspension because the first 10 vol % of fluid would fill the pores and the remaining 30 vol % fluid would separate particles and impart fluidity.

These two examples show a one case where 40 vol % fluid is insufficient to produce fluidity in a 'suspension,' and another case where the suspension at the same solids content would be very fluid. In neither case have we even mentioned chemical additives. Particle packing alone has a large controlling effect on viscous properties.

To take this example one step further, if just a little more fluid is added to the 'suspension' in which particles are touching, the particles will separate a little. Any shear applied to this suspension will tend to produce dilatancy because particles are still really close to one other and collisions and interactions will dominate when the suspension is sheared.

If just a little more carrier fluid is added to the other suspension which is already quite fluid, those particles will separate even further and measured viscosities will decrease as well. With sufficient distance between particles, shear rates will have to be fairly high before particles will begin to interact, collide, and produce any dilatant effects.

These two examples show very clearly how particle physics alone affects the rheological properties of suspensions.

Surface Area

Particle physics directly controls surface areas of powders. As particle sizes decrease, surface areas of particles per unit mass of powder increase substantially. To change the surface area of a powder, one has only to change its particle size distribution. The mass specific surface area (SSA) is the surface area per gram of powder. The volume specific surface area (VSA) is the surface area per cubic centimeter of powder. In industry, mass SSAs of powders and mass %s of chemical additives are commonly used.

Table 15.1 shows the relationships between particle diameter, surface area per particle, number of particles per true cm^3 of powder, and surface area per true cm^3 of powder.

Table 15.1 Relationships between particle size,
numbers of particles, and surface areas (for spherical particles).

Diameter (μm)	Surface Area per particle (m²)	Number of particles per true cm³ of powder	Surface Area per true cm³ of powder (m²)
10 000.	0.31 X 10⁻³	1.9 X 10⁰	0.000 597
1 000.	0.31 X 10⁻⁵	1.9 X 10³	0.005 97
100.	0.31 X 10⁻⁷	1.9 X 10⁶	0.059 7
10.	0.31 X 10⁻⁹	1.9 X 10⁹	0.597
1.	0.31 X 10⁻¹¹	1.9 X 10¹²	5.97
0.1	0.31 X 10⁻¹³	1.9 X 10¹⁵	59.7
0.01	0.31 X 10⁻¹⁵	1.9 X 10¹⁸	597.

Chemical additives react with, or are adsorbed onto, the surfaces of powders. Fixed amounts of additive (added as constant mass % of powder present) will achieve a range of coverages (areal densities) on powder surfaces from batch to batch as the particle size distributions and SSAs vary.

This can be a rather complex relationship because particle size distributions alone affect viscosities and rheologies, and the surface areas present then also affect the coverage and efficiencies of the chemical additives used to adjust the viscosities and rheologies.

To control PSD alone may allow SSAs to vary from batch to batch. To control SSA alone may allow PSDs to vary from batch to batch. Both affect viscous and rheological properties. Optimum performance can be achieved when both PSD and SSA are controlled. This requires very strict, precise controls of one's process. Without such controls, it is easy to have viscous and rheological properties that vary from batch to batch due to particle physics variations.

Surface Properties

One final particle physics property to consider is the nature of the surfaces. Some powders can have relatively smooth surfaces and others can have surfaces that are tortuous. This property is not easily controlled by the end user. If surface texture can be demonstrated as a problem, the solution is to work with the supplier to eliminate surface property variations.

In one plant, dry powders from a particular supplier usually flowed well, but every so often a batch came in where the surfaces were really rough and the powders didn't flow well. The angles of repose of the dry powders in those batches were very different from the normal powders. In most cases, when a beaker full of the powder was poured into a pile on a table, the resulting powder cone was fairly low because the particles flowed well to create a cone with a relatively large diameter base and a low height at the center. Some powder samples, however, produced relatively tall piles with much smaller diameter bases. These powders didn't flow well. Their surfaces were very rough by comparison to the normal powders.

It was determined that when the dry powders didn't flow well, suspensions made with those same powders didn't flow well either, and process yields decreased.

All routine properties of all of these powders, however, were within spec. This problem only became apparent as a test was sought to show differences between the various samples. Following the identification of surface texture as a problem, the angle of repose test was added to the other routine tests. Until this test was identified, it had been apparent that some of the powders behaved differently than the 'normal' powders, and processing was more difficult with those powders, but none of the other routine tests showed any differences between the various powders.

The customer made the supplier aware of this problem, and the supplier solved it. When notified of this problem by the customer, the supplier performed the test, identified the problem, studied their process, and incorporated appropriate changes to prevent the surface roughness problems from recurring.

The interesting aspect of this story is that the customer learned that their 'tight' specifications didn't cover all important properties. When they recognized that the powder surfaces could cause problems, they learned which test would identify this problem, and then they communicated that information to the supplier.

How many companies have problems that come and go (apparently randomly) that are due to materials specifications that don't define all appropriate properties? This is an interesting question, and a difficult problem to solve. It is much easier to identify a property (and a routine test) that is no longer needed to adequately specify the desired properties of a mineral, than to identify a specific property (and a test to characterize it) that is missing from, but should be included in, the specification list.

Each randomly appearing problem could have a single test that can identify, characterize, and track the appearance of that problem. Some such tests could characterize particle physics variations, as in this case. Some such tests could apply to additive chemical controls which will be discussed next.

Additive Chemistry Controls

The utilization of additive chemicals forms the second major category of controls for suspension rheologies. These types of adjustments are usually the final, controlling adjustments made to suspensions. It doesn't matter if particle physics properties are causing problems, or if the interparticle chemistry is out of balance, additives are usually called upon to fix all problems.

This discussion will be split into three categories: pH, deflocculants, and flocculants.

pH

The pH of suspensions is important because each particular powder material will have an isoelectric point at a pH controlled by the powder's composition. The *isoelectric point* is the pH at which the electrostatic surface charges on the clean powder surfaces are zero. Powders typically flocculate at the isoelectric point and they deflocculate as the pH increases or decreases away from that point.

Depending on the natures and concentrations of impurities that travel with the powders, suspension pH can fluctuate from batch to batch. pH is one of the first properties of the interparticle fluid chemistry to check to control suspension rheology.

Many deflocculant chemicals work best at high pH, and many are shipped in solution at high pH. Adding such deflocculants usually causes suspension pH to shift. If the additives to be used in a process work best at a particular pH environment, it may be necessary or advantageous to adjust suspension pH before adding the deflocculants or flocculants.

When two or more additives are combined in a suspension, it is advisable that they be checked for compatibility. The author knows of a process that called for four additives. One of the additives was only supposed to be used at pHs above 8, and another of the additives was only supposed to be used at pHs below 7. This process clearly had potential problems.

Frequently also, additives are the most expensive ingredients in a ceramic body. Chemicals that adjust pH tend to be less expensive than many flocculants and deflocculants. Sometimes the efficiency of the flocculants and deflocculants can be increased simply by adjusting and controlling suspension pH first. If changes to pH allow lower concentrations of other additives to be used to achieve process targets, this should certainly be considered.

For example, adding NaOH to raise pH, and then adding small percentages of deflocculants may be a more cost effective way to produce a better rheological result than adding higher concentrations of the deflocculants without any NaOH additions.

Deflocculants

Many deflocculating chemicals are organic polymers. They are designed to deflocculate by enhancing the electrostatic surface charges on the powders, by providing lubricity during particle/particle collisions, and/or by providing steric layers that help to prevent flocculation.

When organic deflocculants are added to aqueous suspensions, the polymers are usually adsorbed onto the particles by the *hydrophobic effect*. This occurs because the backbones of organic additives are usually *hydrophobic*, which means they don't like water. From an energy point of view, the water tries to force the organic chemicals out of solution by pushing them to an interface such as the suspension surface, or a particle surface. The hydrophobic effect is generally stronger than electrostatic forces. When an additive is electrostatically negative, and the particles are also electrostatically negative, the additives can still adsorb onto and coat particles due to the hydrophobicity of the additives.

When anionic deflocculants adsorb onto electrostatically positive surfaces, the net positive surface charges are quickly cancelled and the surfaces become electrostatically negative. Depending on the structure and composition of the deflocculants, the hydrophobic effect can cause anionic deflocculants to adsorb onto electrostatically negative surfaces and enhance their negative charge densities.

Sodium silicate is the primary member of the category of deflocculants known as inorganic deflocculants. It is a soluble, inorganic additive that functions as a deflocculant primarily by retiring soluble flocculating cations such as Mg^{++}, Ca^{++}, and Al^{+++}.

Organic and inorganic flocculants function differently and they produce different suspension properties. When high levels of flocculating cations are present in suspension, sodium silicate can combine with them to produce insoluble silicates which precipitate and are rendered neutral. Once the flocculating cations have been precipitated as insoluble silicates, they are no longer available to influence suspension electrostatic properties.

Organic deflocculants typically don't retire the flocculating cations (unless they're specifically labeled as chelating agents.) Organic deflocculants may tie up such cations, or mask their effects by coating them (as a coat of paint hides the natural color of wood), but under the influence of high intensity dispersion (HID), the flocculating cations and the deflocculating polymers can all once again be freed into the interparticle soup where they are again free to adsorb onto particle surfaces and generally move suspension properties towards equilibrium when the HID conditions are removed.

When flocculating cations are precipitated as insoluble silicates, HID cannot redissolve them. When flocculating cations are removed by inorganic deflocculants, such as sodium silicate, they are no longer available to affect the interparticle electrostatic charge balance. Once they're removed by precipitation, they will not return. This characteristic forms one of the major differences between the organic and inorganic additives.

These examples show that the two different types of deflocculants behave in fundamentally different ways. In relatively quiescent systems, those differences may never appear. But when the suspensions are exposed to high intensity shear conditions, those differences may become readily apparent.

Another concern when HID is used during processing is that it can break the long chain organic polymeric deflocculants into smaller chains. Many of the organic deflocculants are ionizable polymers. Most such deflocculants are produced in a variety of chain lengths (a variety of molecular weights). Each process that uses such deflocculants usually requires a very specific average chain length for optimum effectiveness of the deflocculant. When organic deflocculants are selected and used at their optimum chain lengths, HID conditions can break the chains, change their average lengths, and render the deflocculants less effective.

Deflocculants should be efficient at concentrations of a few tenths of a percent. Generally speaking, if additions of one or two percent are required, the additive should be considered inefficient and another should be sought. There are exceptions to this, of course. The nature of the additive, the function it is expected to perform, the type of process in which it will be used, and the shear environment to which it will be exposed, should all be factored into the decision whether an additive is cost effective or not.

Flocculants

Many flocculants tend to be soluble, inorganic salts of divalent cations such as Ca^{++} and Mg^{++}. These ions flocculate by cancelling negative electrostatic surface charges and allowing the weak van der Waals forces of attraction to dominate.

Because these flocculants function in the electrostatic realm, one has to pay attention to interparticle fluid conductivities. Electrostatic attractions and repulsions between particles work well in environments where fluid conductivities are low. As fluid conductivities rise, electrostatic forces have less and less influence on suspension properties.

This is an important point because some may consider alternating additions of sodium silicate (deflocculant) and calcium chloride (flocculant) to cancel one another. If too much deflocculant has been added, cancel it by adding more flocculant, and vice versa.

But when the equations for this reaction are examined, the calcium may actually cancel (and precipitate) the silicate, and vice versa, but the remaining ions, Na^+ and Cl^-, will act to increase the interparticle fluid's conductivity. The effectiveness of the particles' electrostatic charges will quickly decrease and disappear as Na^+ and Cl^- concentrations, and conductivities, rise.

As interparticle fluid conductivity increases, additives lose their effectiveness and particles become flocculated due to van der Waals forces. When this occurs, further additions of deflocculants and flocculants all cause suspension viscosities to rise.

High Additive Concentrations

A point needs to be reinforced regarding high concentrations of deflocculants and flocculants. High concentrations of either category of additives can act like the opposite category. High concentrations of flocculants can deflocculate and high concentrations of deflocculants can flocculate.

High concentrations can result from many small chemical additions over time, which bring the overall concentrations up to relatively high values. High concentrations can also be produced by quickly pouring a container of additive into a large tank. If additives are not well dispersed, local regions can contain relatively high concentrations, while the bulk of the suspension sees little, if any, additive.

When possible, chemical additives should be mixed into suspensions in a blunger tank. The agitators in holding tanks are not designed to disperse chemicals, but to provide recirculation to prevent settling. Unfortunately, many additives are introduced into suspensions in holding tanks. Under such circumstances, it is easy to produce local regions with relatively high concentrations of additives which may take many hours to disperse thoroughly throughout the whole tank. Regions with locally concentrated additives can form 'rocks' which sink to the bottom of the tank and never disperse. This is especially the case in holding tanks with very mild agitation.

If it is possible to recirculate a suspension through a continuous high intensity disperser (CHID) and back to the tank, the best place to add chemical additives is at a point in the piping at the entrance to the CHID unit. Not only does this disperse additives well, but it does it quickly.

Aging Processes

Depending upon the level of agitation and the nature of the powder ingredients, aging may be required to achieve suspension stability. When relatively low intensity agitators are used in blungers and holding tanks, suspensions may require several days to reach equilibrium. Such suspensions are frequently characterized by holding tank viscosities that drift to higher or lower values from day to day. When more additives are introduced into the holding tanks to help stabilize such suspensions, viscosities may continue to drift as the suspension moves towards equilibrium over the next several days.

Alternatively, when high intensity dispersion (HID) is used during the batching process, the necessity for aging can be minimized. If suspensions are beat up severely in the main blunger (more severely than they'll ever see anywhere else in the process) so all particles report as individuals, the final equilibrium state of the suspension can be achieved relatively quickly.

There appears to be a philosophy out there which calls for as little agitation as possible to mix a batch. Then, this philosophy also calls for as little agitation as possible in the holding tanks so as not to disturb the suspension properties (especially viscosities) during the suspension's short residence time until it can be sent to the process. This philosophy generally doesn't work. If such limited agitation levels are actually achieved, suspensions will not be sufficiently mixed to achieve uniformity nor stability.

If enough agitation is present in the main blunger to start the mixing process, and holding tank agitation is sufficient to recirculate suspensions properly, aging can take place in which particles will slowly be freed to report as individuals. As this happens, viscosity will constantly be drifting because newly freed particles will move into the interparticle fluid environment and the fundamental nature of the suspension will change from day to day. This aging process can take days to weeks to run to completion.

High intensity dispersion conditions can impose enormous amounts of mechanical dispersion energy on a suspension in a short period of time. HID conditions are defined as agitator tip speeds in excess of 5000ft/min. A 4" diameter impeller running at ~5000 rpm produces HID conditions sufficient to mix a 5 gallon container of suspension in the lab. Upon exposure to HID conditions, room temperature suspensions quickly rise in temperature beyond 70°C. High suspension temperatures at the completion of HID is an excellent indicator that it has been successful.

Production HID systems require large motors to produce very intense conditions which can substitute for several days of aging.

Partially Soluble Ingredients

Some powders used in ceramic forming bodies are partially soluble. Even when levels of solubility are small, soluble cations can enter solution over periods of days to cause constant viscosity changes.

This can happen in both flocculating and deflocculating directions. Dolomite, for example, puts calcium and magnesium ions slowly into solution to cause bodies to flocculate with time. Nepheline syenite puts sodium ions slowly into solution to cause bodies to deflocculate with time. One solution to such problems is to replace these materials (if possible) with other raw materials. When such materials must be used, frequent adjustments with additive chemicals may be required to maintain desired viscous and rheological properties.

Sometimes, and this may require some luck and/or serendipity, two materials may be found that balance each other. If the flocculating effects of the ions dissolving from one mineral are counterbalanced by the deflocculating effects of the ions dissolving from another mineral, and the dissolution rates of the two are about the same, a win-win situation may appear.

One needs to pay attention to the solubility of all raw materials. Most ceramic raw materials are insoluble in water and that is desirable when they are to be used in aqueous suspensions. When partially soluble materials must be used, pay attention to that fact, and be prepared to make adjustments to production suspensions on a daily basis.

Attacking the Symptoms or the Cause?

When suspension rheologies and/or viscosities need adjusting, it is handy to know the source of the problem that caused the viscosity drift. When one knows the source of the problem, it can be fixed.

High suspension viscosities could be the result of improper particle physics controls, or improper additive chemistry controls. The typical production solution is to fix the problem by adjusting viscosities with chemical additives. Chemical additives **can not** always fix production viscosities or rheologies caused by particle physics problems.

In a daily batch, if the particle size distribution narrows, packs poorly, and causes the viscosity to increase, deflocculants may be able to reduce viscosity to appropriate levels, but the rheology will become more dilatant as the concentration of deflocculants increases. In the next daily batch, if the particle size distribution broadens a little, packs better, and the viscosity decreases, flocculants may be able to raise viscosities to appropriate levels, but the rheology will become less dilatant and more shear-thinning with the addition of the flocculants.

These two cases can be produced by fluctuations in particle size distribution from one day to another. The result can be a more dilatant suspension one day and a more shear-thinning suspension the next day. Note that when the suspensions in this example are adjusted to constant production viscosities at a particular shear rate, their production rheologies change.

The point of this example is to show that viscosity variations in a production suspension, even when caused by particle physics fluctuations, are usually corrected using additive chemistry. This needs to be repeated: Even when **particle physics fluctuations** cause the viscous and rheological problems, suspension properties are usually adjusted using **additive chemistry**. Then, when **chemical fluctuations** cause the viscous and rheological problems, suspension properties are usually adjusted using **additive chemistry**.

Most attempts to correct suspension viscosity problems, regardless of the cause, are made using additive chemistry adjustments. This type of solution frequently attacks the symptoms, without making any real changes to the underlying problem. This is okay for a temporary fix, but the long-term solution requires the source of the problem to be identified. With the problem's source identified, appropriate corrections can be applied.

Summary

Both particle physics and additive chemistry have major effects on suspension viscosities and rheologies. When a production suspension's viscosity or rheology is out of spec, an attempt should be made to determine the cause. Too often the symptom (out of spec viscosity or rheology) is attacked with an additive chemical adjustment, while the underlying problem is overlooked.

Improper control of either particle physics or additive chemistries in suspensions can produce disastrous results. An ideal interparticle chemistry can be ruined by improper particle physics, and ideal particle physics properties can be ruined by improperly adjusted chemistries. Both particle physics and interparticle chemistry must be addressed and optimized to achieve optimum viscosity and rheology in production suspensions.

References

1. <u>Webster's Seventh New Collegiate Dictionary</u>, G. & C. Merriam Company, Springfield, MA (1965).

2. Einstein, A., <u>Investigation of the Brownian Movement</u>, Dover, NY (1956).

3. Hafaiedh, A., "Computer Modelling of the Rheology of Particulate Suspensions," Alfred University, PhD Thesis (1988).

4. Bedeaux, D., "The Effective Viscosity for a Suspension of Spheres," *J. Coll. Int. Sci.*, <u>118</u> 80-90 (1987).

5. Robinson, J.E., "The Viscosity of Suspensions of Spheres," *J. Phys. & Coll. Chem.*, <u>53</u> 1042-1047 (1949).

6. Everson, G.F., <u>Rheology of Disperse Systems</u>, pp. 61, Pergamon Press, London (1959).

7. Funk, J.E., and Dinger, D.R., <u>Predictive Process Control of Crowded Particulate Suspensions Applied to Ceramic Manufacturing</u>, Kluwer Academic Publishers, Boston, MA, pp. 452-453 (1994).

8. Funk, Professor James E., private communications.

9. Funk, J.E., and Dinger, D.R., *op. cit.*, 699-710.

10. Funk, J.E., and Dinger, D.R., *op. cit.*, 37-120.

Glossary

3-D Gel Structure - The 3-D gel structure is the complete structure that forms throughout the whole volume of a suspension as particles flocculate. Given sufficient time, all particles in a flocculated suspension will be immobilized into this structure.

Additive Chemistry - The complete additive chemistry package includes all of the various chemicals (flocculants, deflocculants, binders, etc.) that are added to suspensions to control their properties.

Aging Process - When a suspension is stored to allow all constituents time to come to equilibrium, this is considered to be the aging process. Delamination, deagglomeration, and dynamic positioning of all ions and chemicals in the interparticle fluid can occur. All such changes move in the general direction of an equilibrium system in which suspension properties are stable with time. This occurs over a period of several days depending on batching procedures, and blunger and storage tank impeller intensities.

Anionic Polyelectrolyte - Many organic additives (deflocculants) have polymeric structures with many ionizable cations along their lengths. When the cations ionize and move off into solution, the structure that remains is a long chain organic with many negatively charged sites along its length. These types of additives are known as anionic polyelectrolytes.

Apparent Viscosity - The apparent viscosity of a suspension is the ratio of the measured shear stress to the imposed shear rate. Under each different set of shear conditions, a non-Newtonian suspension will have a different apparent viscosity. At each set of measurement conditions, one could say, "The suspension *appears* to have a particular viscosity." That viscosity is the *apparent* viscosity. The apparent viscosity must be accompanied by the shear rate at which it was measured. To say that a suspension has an apparent viscosity of 1000 mPa-s is meaningless. It becomes meaningful when the shear rate of the measurement conditions are also specified. For example, "The apparent viscosity of the slip is 1000 mPa-s at $100s^{-1}$," is a valid statement and a useful piece of information.

Bingham Rheology - The Bingham form of rheology is characterized by a linear behavior of shear stress vs shear rate after the shear stress has exceeded the yield stress value. Mathematically, Bingham rheology is Newtonian behavior plus a yield stress.

Capillary Viscometer - A capillary viscometer measures viscosities by measuring the pressure drops required when fluids are pumped through long, small diameter tubes (capillary tubes).

Colloids - Colloidal particles are defined as those with effective diameters less than ~1mm.

Cone-and-Plate Viscometer - This type of rotational viscometer measures the viscosity of fluids in the gap between an inverted cone spinning on a stationary plate. Included angles between the cones and the plates can be very small (fractions of a degree), so sample volumes are small. High shear conditions can be easily achieved in cone-and-plate viscometers.

Continuous High Intensity Dispersion (CHID) - This type of device subjects suspensions to HID conditions as they flow continuously through it. Typically, one pumps suspension into a CHID and the overflow returns to the process tank. See also **High Intensity Dispersion**.

Cup-and-Bob Viscometer - This type of rotational viscometer measures viscosities of fluids as they are sheared in the gap between a rotating bob and a stationary cup (or between a rotating cup and a stationary bob.) With small gap sizes and high rpms, relatively high shear conditions can be achieved in this type of viscometer.

Deflocculant - This category of chemicals includes both organic and inorganic additives that deflocculate a body.

Deflocculate - To deflocculate a suspension, one sets conditions (using deflocculant additives) that cause all particles to repel each other. When particles stay as far apart as possible, they will travel as individuals, and suspension viscosities will be relatively low. Gelation will not usually occur in well-deflocculated suspensions.

Dilatant Blockage - When particles in suspension collide due to dilatancy, and the dilatancy is severe enough to cause the particles to be mechanically interlocked into a relatively dry, porous compact that blocks a flow channel, and thereby causes flow to cease, the compact is known as a dilatant blockage.

Dilatant Rheology - This form of rheology is also known as shear-thickening rheology. Measured viscosities increase as shear increases due to dilatancy. During flow, particles collide and as those collisions intensify as shear rates increase, apparent viscosities increase as well.

Dynamic Viscosity - The dynamic viscosity is the ratio of shear stress to shear rate of a flowing fluid.

Floc - A floc is a weakly bonded group of particles. Gelation (flocculation) pulls the particles together to form flocs and the shear from flowing fluids can break them apart again. In high solids suspensions, flocs are the first step in the gelation process. Small flocs form, grow into large flocs, and they all then combine to form large 3-D gel structures.

Flocculant - A flocculant is an additive that causes particles in suspension to be attracted to one another and to flocculate.

Flocculate - When particles flocculate, they are attracted to one another by van der Waals forces and they thereby form weakly bonded structures known as flocs. In suspensions, flocs of particles can travel together until they are broken by shear forces. When shear decreases, the attractive forces that are always present pull the particles towards one another and they once again flocculate into groups of particles. In quiescent high solids suspensions, flocculation pulls all particles and flocs into large, continuous, 3-D gel structures that extend throughout the whole volume of suspension.

Gel Structure - See **3-D Gel Structure**.

High Intensity Dispersion (HID) - High intensity dispersion conditions are defined as blunging with impeller tips speeds of 5000ft/min or greater.

Hydrophobic Effect - The hydrophobic effect occurs when water pushes hydrophobic organic chemicals to the nearest surface (e.g., the interface between the fluid and the atmosphere, or between the fluid and a particle). Thermodynamically, water molecules prefer to be near other water molecules. As a result, they try to minimize their contact with hydrophobic organic chemicals. This effect is stronger than electrostatic repulsive forces, so negatively charged organic additives can be forced onto negatively charged particle surfaces.

Hysteresis - In a rheogram, hysteresis occurs when the rheogram trace from the viscometer during acceleration through a range of shear rates differs from the trace from the viscometer during deceleration through that same range of shear rates. It is an indication that the suspension being measured exhibits thixotropic or rheopectic time-dependent character.

Infinite-Sea Viscometer - This is a rotational viscometer in which a rotating bob can make a measurement as it is immersed in a relatively large container of fluid or suspension. The container that holds the fluid or suspension can be relatively large, as the name implies.

InterParticle Spacing (IPS) - The IPS is the average distance between particles in a suspension. Generally, as the IPS increases, the measured viscosity will decrease.

IsoElectric Point (IEP) - The IEP of a mineral is the pH at which a suspension with clean particle surfaces exhibits zero surface charge. *Clean* particle surfaces are those which are not covered by additives which can alter the measured surface charges. Generally, particles will exhibit positive surface charges at pH values less than the IEP and negative surface charges at pH values greater than the IEP.

Newtonian Rheology - Simple fluids, each of which have a single, characteristic viscosity regardless of the applied shear conditions, are examples of Newtonian fluids. Newtonian rheology is characterized by linear shear stress versus shear rate rheograms that begin at the origin. Newtonian fluids do not have yield stresses. Because a single Newtonian viscosity characterizes each simple fluid, the shear rate at which the viscosity measurement is made is unimportant.

Non-Newtonian Rheology - All fluids that do not exhibit Newtonian rheology are considered to be non-Newtonian. The apparent viscosities of all non-Newtonian fluids vary as the applied shear rates, or time of application of shear, varies. Non-Newtonian rheologies include dilatant, pseudoplastic, Bingham, yield-dilatant, yield-pseudoplastic, thixotropic, and rheopectic rheologies.

Onset of Dilatancy - The onset of dilatancy is the shear rate at which a suspension begins to exhibit dilatant character. At all shear rates higher than the onset of dilatancy, suspensions will exhibit dilatant rheological properties.

Oscillating Viscometer - These types of viscometers use oscillating probes to measure viscous properties. When these probes are immersed in suspensions and fluids, their electronic controllers drive their frequencies and amplitudes of oscillation. The damping effects of the suspensions and fluids on the oscillating probes allow viscosities to be calculated.

Particle Physics - All properties of powders and suspensions that relate to physical properties, such as particle size distribution, surface area, packing, surface roughness, etc., are considered particle physics properties.

Pseudoplastic Rheology - Also known as shear-thinning rheology, pseudoplastic rheologies decrease in viscosity as applied shear rates increase. A pseudoplastic suspension flowing in a pipe will have a lower apparent viscosity when it is flowing fast, than it does at a lower flow velocity.

Rheogram - Rheograms are plots of shear stress versus shear rate, apparent viscosity versus shear rate, shear stress versus time, or apparent viscosity versus time, for fluids, suspensions, and forming bodies. They are used to characterize the rheologies (the apparent viscosity behaviors) of suspensions as functions of shear rate and time.

Rheology - Rheology is the study of the viscous behaviors of fluids and suspensions as functions of shear rate and time. Simple fluids are each characterized by a single viscosity. But there are many fluids and suspensions characterized by viscosities that vary as they are sheared at different shear rates and for different time durations. Rheology quantifies the various types of viscosity behaviors.

Rheometer - Just as a viscometer is used to measure viscosities, a rheometer is used to measure rheologies. For a viscometer to be described as a rheometer, it must be capable of measuring viscosities as functions of shear rate and/or time.

Rheopexy - This form of rheology is characterized by the increase of measured viscosities with time at constant shear rate.

Shear Rate - The shear rate is the velocity gradient imposed upon a fluid or suspension during shear. It has units of velocity/time, such as (cm/sec)/cm, which are usually simplified to reciprocal seconds, 1/sec, or simply s^{-1}.

Shear Stress - The shear stress is the shear force per application area required to shear a fluid or suspension at a particular rate. To measure viscosities, many viscometers impose a shear rate on a fluid and they then measure the resulting shear stresses. Typical engineering units for shear stress are *psi*. The Pascal, *Pa*, is the typical unit within SI systems.

Shear-Thickening Rheology - Fluids which exhibit increasing apparent viscosities as shear rates increase are examples of shear-thickening rheology. See **Dilatant Rheology**.

Shear-Thinning Rheology - Fluids which exhibit decreasing apparent viscosities as shear rates increase are examples of shear-thinning rheology. See **Pseudoplastic Rheology**.

Shear*Time History - Thixotropy and rheopexy respond to intensities of imposed shear and the durations of the exposures. As such, they are known as time-dependent rheologies. The shear*time history is a way to quantify both the intensity and duration of shear exposure. It is the area under the shear rate versus time curve.

Slip - A slip is a ceramic suspension that contains several body ingredients. When all ingredients of a body are present, the suspension can be described as a *body slip*.

Slurry - A slurry is a suspension that contains a single mineral ingredient. Ball clay slurries, kaolin slurries, and coal slurries are examples of suspensions containing only ball clay, only kaolin, and only coal, respectively.

Steric Hindrance - The word *steric* suggests a spatial phenomena. Steric hindrance occurs when particles are coated with additives and they can come no closer to each other than the additive thicknesses allow. The particles are hindered, spatially, by their coatings, from coming close to one another. Even if the coatings on each particle touch one another, the particle surfaces are still separated by a distance equal to the sum of the two coating thicknesses. This is steric hindrance.

Syneresis - Syneresis occurs during states of over-flocculation. Not only do gel structures form, but attractive forces are so strong in syneretic systems that they densify the structures with time and expel interparticle fluid to the gel surfaces.

Thixotropy - This form of rheology is characterized by the decrease of measured viscosities with time at constant shear rate.

Time-Dependent Rheology - The two time-dependent rheologies, thixotropy and rheopexy, respond to imposed shear rates and time. Final viscosities are not achieved instantaneously, but they require time at constant shear rate for the suspensions to come to equilibrium conditions (and relatively stable constant viscosities.)

Time-Independent Rheology - The six time-independent rheologies, pseudoplastic, Newtonian, dilatant, yield-pseudoplastic, Bingham, and yield-dilatant, occur independent of time. These behaviors are not dependent on time, so their effects should be visible as soon as a particular shear rate is imposed.

Viscometer - A viscometer is an instrument that can measure viscosity.

Viscosity - Viscosity is defined as the ratio of shear stress to shear rate that characterizes the behavior of fluids and suspensions. High viscosity fluids are relatively thick, slow flowing fluids such as molasses. Low viscosity fluids are relatively thin, fast flowing fluids such as water.

Yield Stress - The yield stress value characterizes the strength of the gel structure of a quiescent suspension. The yield stress of a suspension is the stress that must be exceeded before flow will occur. Before flow has begun, any applied stresses less than the yield stress may cause elastic deformation to the gel structure, but no flow, nor rearrangement of suspended particles or fluid molecules, will occur. Once stresses exceed the yield stress and flow occurs, applied stresses may actually fall below the yield stress value. When flow stops, gelation will rebuild the structure and the yield stress will return. Without a yield stress, a formed ceramic ware would not be able to hold its shape.

Yield-Dilatant Rheology - This is a dilatant (shear-thickening) rheology that has a yield stress. This is the one, truly important rheology within ceramic process systems. See **Dilatant Rheology**.

Yield-Pseudoplastic Rheology - This is a pseudoplastic (shear-thinning) rheology that has a yield stress. See **Pseudoplastic Rheology**.

Index

181

www.ingramcontent.com/pod-product-compliance
Lightning Source LLC
Chambersburg PA
CBHW071425180526
45170CB00001B/222